Nuclear Safety
Risks and Regulation

William C. Wood

American Enterprise Institute for Public Policy Research
Washington and London

William C. Wood is assistant professor of economics at the University of Virginia and a former assistant program analyst for the U.S. Nuclear Regulatory Commission. He is the author of *Insuring Nuclear Power* and several economic and technical articles on nuclear safety.

I would like to thank Roger Sherman, Linda Cohen, Joel Yellin, John Sullivan, and Peter Bloch for helpful comments on an earlier version of this manuscript.

W.C.W.

Library of Congress Cataloging in Publication Data

Wood, William C.
    Nuclear safety regulation.

    (AEI studies ; 370)
    1. Atomic power—Law and legislation—United States.
2. Atomic power-plants—Safety regulations—United
States. I. Title. II. Series.
KF2138.W66   1983   343.73'0925   82-22730
ISBN 0-8447-3508-6   347.303925

AEI Studies 370

The American Enterprise Institute for Public Policy Research, established in 1943, is a publicly supported, nonpartisan, research and educational organization. Its purpose is to assist policy makers, scholars, businessmen, the press, and the public by providing objective analysis of national and international issues. Views expressed in the institute's publications are those of the authors and do not necessarily reflect the views of the staff, advisory panels, officers, or trustees of AEI.

Contents

1

Making Decisions on Nuclear Safety

Nuclear technology as a source of energy in the United States was clearly in trouble in the early 1980s. Electric utilities had turned away from nuclear reactors for new generating capacity, and some were cancelling nuclear projects that had already been started. Manufacturers of nuclear power plants, receiving no new orders, were producing only to fill the existing orders and were phasing down their nuclear research and development. Concerns over nuclear safety persisted. Symbolizing the industry's difficulties, the disabled Unit 2 reactor at Three Mile Island was slowly being cleaned up from the 1979 accident as its corporate owner worked to avoid bankruptcy.

Such was the state of a technology that initially had promised electricity too cheap to meter and later had offered at least the prospect of reduced dependence on fossil fuels. Were nuclear energy's difficulties inherent in the technology and only catching up with events in the 1980s? Or was the technology sound and its difficulties the result of faulty institutions? This work examines safety regulation as an institution that may well have been a significant contributor to the difficulties of nuclear technology. In particular, the results of regulation by the Nuclear Regulatory Commission (NRC) since it was set up as an independent agency in 1975 are explored.

This chapter provides an overview of nuclear regulation, followed by an account of the legislative rationale and the economic rationale for safety regulation of nuclear technology. The chapter concludes with a discussion of the conflict between the economic and legislative rationales. Chapter 2 reviews the NRC's structure and safety programs and shows how many of its problems are unlike those facing other regulatory agencies. Chapter 3 assesses the results of nuclear regulation. It examines whether nuclear regulation has produced an acceptable level of safety and the evidence on acceptability. It concludes that the existing level of safety is not being achieved cost effectively and tries to find the sources of regulatory failure. Chapter 4 evaluates proposed changes in nuclear safety regulation. Chapter 5 contains a summary and conclusions.

1

Overview of Nuclear Regulation

The Technology. It is sometimes helpful, and sometimes misleading, to characterize the nuclear reactor as "a better way to boil water." This characterization is helpful in that it indicates the way the nuclear reaction produces electrical energy: the fission, or splitting, of uranium atoms results in great releases of heat, which produce steam. The steam produces electricity in a turbine generator, just as steam produced by conventional fuels turns a turbine generator. Thus it is not exotic technology that produces "nuclear" electricity; only the source of heat is fundamentally different from that in conventional plants.

It is, however, misleading in some contexts to think of the nuclear reactor as a better way to boil water. Such a narrow view tends to obscure the fact that the nuclear reactor is only part of a complex system that extends from the mining of uranium to the disposal of reactors' radioactive wastes. Seeing nuclear technology as an alternative heat source that can readily be "plugged in" to existing technologies understates the technical and institutional difficulties of its use.

The distinctive safety difficulties of nuclear power stem from radioactivity. The same forces that release the energy in the nuclear reactor are the forces that can lead to sickness, genetic defects, or death. Thus radioactivity in all its aspects must be contained: the low-level radioactivity of normal operation, the potentially massive releases of radioactivity that could result from accidents, and the long-term radioactivity of the wastes generated by normal operation. Any human undertaking involves risk, and the risk in nuclear power results from uncertainty over the effects of radioactivity and the ability to contain it. The potential harm of radioactivity puts a strain on institutions ranging from insurance to law enforcement that is not present with other technologies. It can, therefore, be quite misleading to think of nuclear energy just as a better way to boil water.

The Industry. The nuclear industry in the United States consists of the reactor manufacturers, the architect-engineers who build nuclear plants, the utilities that operate them, and a host of smaller firms that provide components and services. There are today only four domestic reactor vendors: Westinghouse, General Electric, Combustion Engineering, and Babcock and Wilcox. Westinghouse and General Electric are the larger and older vendors, together accounting for more than two-thirds of the installed nuclear capacity. Westinghouse and the two smaller companies produce the pressurized water reactor (PWR), which Westinghouse developed as a partner in the navy's research program in nuclear propulsion. General Electric produces the boiling water reactor (BWR) design which it developed. While there are great engineering differences between reactors depending on whether the water covering the hot nuclear fuel is intended to remain a pressurized liquid or to boil, the policy problems for PWRs and BWRs are the same. Both designs pose the possibility of radioactive releases to the environment.

The four reactor manufacturers supply the heart of the nuclear power plant, the nuclear steam supply system. The rest of the plant is designed, and the facility is constructed, by an architect-engineering firm. There is some vertical integration of the architect-engineering function; some early plants were built entirely by the reactor vendor, and other plants were built by the operating utility. For the most part, however, the plant is built by a firm separate from both the reactor vendor and the operating utility.

The utilities that operate nuclear power plants are generally the larger private and public electric utilities. Smaller utilities often have had a share of ownership, at first because of antitrust requirements and more recently because of larger utilities' excess capacity in the face of slowed growth of demand. To the operating utility, an on-line nuclear power plant is much like any other plant: its output is sold to produce revenue, its operating expenses are passed through to consumers, and a rate of return is permitted on the investment. The rate of return is determined by state regulators who may act independently of federal regulators. Compared with the other plants in a utility's system, nuclear plants have a greater capital investment but lower operating costs. It is, therefore, expensive to have nuclear plants sitting idle, and utilities run them around the clock during normal operations.

Rounding out the nuclear industry are scores of smaller firms that supply components, services, and technical help to the other firms. Despite the role that diverse firms have in the construction and operation of nuclear plants, the NRC holds the operating utility responsible for the safety of the plant.[1] When a "licensee" is referred to in this study, the term will mean the operating utility.

The Nuclear Regulatory Commission. The federal agency that regulates civilian use of nuclear technology is the Nuclear Regulatory

1. For a restatement of this NRC policy, see U.S. Nuclear Regulatory Commission, "NRC Views and Analysis of the Recommendations of the President's Commission on the Accident at Three Mile Island," NUREG-0632 (Washington, D.C.: U.S. Nuclear Regulatory Commission, November 1979), p. B-1.

Commission. Its jurisdiction includes medical and industrial uses of nuclear materials as well as nuclear reactors and the nuclear fuel cycle from milling to waste disposal.

The Energy Reorganization Act of 1974 separated the NRC from its parent agency, the Atomic Energy Commission (AEC).[2] The AEC had been responsible for promoting and regulating nuclear technology. The 1974 reorganization set up the NRC strictly as a regulatory body and transferred the AEC's promotional responsibilities to the Energy Research and Development Administration (later part of the Department of Energy).

The Nuclear Regulatory Commission is organized like other independent regulatory commissions, with commissioners appointed by the president and confirmed by the Senate. One commissioner is designated chairman of the agency by the president. The five commissioners serve staggered five-year terms, and no more than three may belong to the same political party. The NRC staff numbered 3,139 as of fiscal year 1981, when the authorized budget was $449 million.

The 1974 act setting up the NRC established three program offices within the new agency. The Office of Nuclear Reactor Regulation was set up to license and regulate nuclear reactors. The Office of Nuclear Material Safety and Safeguards was set up to oversee the use of nuclear materials, including in particular the nuclear fuel cycle. The Office of Nuclear Regulatory Research was set up to do research on nuclear safety matters at the commission's direction.

Two other program offices were organized by the new commission: the Office of Inspection and Enforcement to ensure compliance and the Office of Standards Development to issue regulatory standards. The standards office was merged with the research office in 1981. A small staff reports directly to the commission, and nine staff offices report to the executive director for operations, a sort of general manager who serves at the pleasure of the commission. Panels for hearing individual licensing actions and appeals are separate from the commission, and the Advisory Committee on Reactor Safeguards (ACRS), made up of scientists who serve part-time, conducts independent reviews.

The NRC has broad authority to regulate nuclear technology. Regulatory policy is spelled out both in proceedings that apply to all plants and in individual licensing decisions.

The Critics and Intervenors. No introduction to nuclear regulation would be complete without some description of the critics of nuclear

2. P.L. 93-438, 42 USC 5801 (1974).

4

power and of the intervenors in individual licensing actions. These diverse bodies range from loosely knit groups of protestors in the areas of nuclear sites to the technically sophisticated Union of Concerned Scientists, a Cambridge-based organization whose ranks include former industry and NRC engineers who resigned over safety issues. Strategies successfully employed by antinuclear groups range from simple stalling tactics to the raising of genuine safety issues not previously recognized by the industry or by the NRC.

To say that the future of nuclear power in the United States is uncertain is a great understatement. Nuclear critics sense a victory in the de facto moratorium on new orders that has prevailed since before the Three Mile Island accident. Nuclear proponents foresee a time in which the nation will badly need the power that might have been supplied by nuclear plants, had that option not been foreclosed by inept regulation and policy. To assess the future of nuclear technology and nuclear regulation, one must look at the legislative and economic context of regulation.

The Legislative Setting of Nuclear Regulation

Nuclear technology was a direct byproduct of the World War II military effort, and even today civilian nuclear power may suffer from an association with the same forces that caused the destruction at Hiroshima and Nagasaki. Because nuclear technology was a highly guarded secret at the end of the war, it is not surprising that the government monopoly on the new technology was continued under the Atomic Energy Act of 1946.[3] Because of the preoccupation with finding highly visible peaceful uses of atomic energy, it is also not surprising that the 1946 act makes no substantive statement on public health and safety.

By 1954, sentiment was widespread that both government and industry should undertake the development of nuclear applications. The Atomic Energy Act of 1954 was enacted, setting up "a program to encourage widespread participation in the development and utilization of atomic energy for peaceful purposes to the maximum extent consistent with the common defense and security and with the health and safety of the public."[4] In regulating the yet-to-be-established private nuclear industry, the AEC was to "protect health and minimize danger to life and property," yet Congress gave the AEC no

3. See Harold P. Green and Alan Rosenthal, *Government of the Atom: The Integration of Powers* (New York: Atherton, 1963), pp. 2-5, on this period.
4. See the 1954 act as amended, P.L. 83-703, chap. 1, sec. 3(d).

guidance on just how safe nuclear technology had to be.[5] In the hearings on the 1954 act, there was no direct consideration of what the possible safety hazards involved might be.[6]

Because the Energy Reorganization Act of 1974 separated the NRC from the AEC to enhance safety through an independent regulatory body, it might be thought that Congress explicitly laid out a standard of safety or at least put forth some rationale for safety regulation in 1974. The text of the act, however, contains no such material. It only reaffirms that the NRC should work "to assure public health and safety," and it requires the NRC to report annually on nuclear safety.[7] Debates over the bill took place while Congress was preoccupied with Watergate and the Arab oil embargo of 1973–1974, so the rationale for safety regulation and the legislative view on safety received little attention.[8] Following the 1974 act, the NRC was not sure whether it was required or allowed to consider the costs of its regulations, or whether it had to pursue nuclear safety regardless of the cost. Nor was it sure whether Congress wanted it to ensure that its regulations were not so severe as to prevent the existence of a domestic nuclear industry.

As this quick survey indicates, Congress has issued no clear statement of the rationale for government regulation of nuclear power. Nor is there a direct mandate to achieve a particular safety level, even a qualitative mandate. The legislative rationale must, therefore, be inferred. From a reading of the legislative history, three possible legislative rationales emerge:

1. National security. Since nuclear technology involved secrets important to national security, it was important to make sure that information did not get into the wrong hands. Several parts of the 1954 act were aimed at preserving the secrecy of nuclear technology, with private firms given access on a "need-to-know" basis.[9] This argument, however, provides a rationale only for regulating the information flow and not for the commission machinery for administering safety regulation. In any event, nuclear technology was widely known

5. See Elizabeth S. Rolph, *Nuclear Power and the Public Safety: A Study in Regulation* (Lexington, Mass.: Lexington Books, 1979), p. 28, citing P.L. 83-703, 1954, chap. 1.

6. Ibid.

7. P.L. 93-438, sec. 307(c).

8. The House passed the bill December 19, 1973, and the Senate passed it August 15, 1974, in slightly different form. Conference reports were agreed to October 9 and 10, and President Ford signed the measure October 11, 1974.

9. On security and secrecy, see Green and Rosenthal, *Government of the Atom*, pp. 199-201.

by 1974, and secrecy of information surely played little part in the Energy Reorganization Act of that year.

2. "It's dangerous, so regulate it." In some legislators' minds, the fact that nuclear technology is dangerous was sufficient reason for government regulation. The 1974 act was passed when regulation of the workplace and of consumer products needed no legislative foundation more compelling than a perceived lack of safety.[10] In this setting, safety was seen as an "all-or-nothing" quality; either something was safe or it was not. Obviously Congress wanted nuclear plants that were safe rather than unsafe. Such loose thinking about safety was promoted by early pronouncements of the AEC that nuclear power was "safe" and by early thinking that there were pervasive "threshold" effects in nuclear safety. Early work indicated, for example, that there was a threshold dose of radiation below which living cells would repair themselves and be none the worse for the wear.[11] Radiation doses below the threshold would be safe. Later work indicating the lack of a threshold might have forced Congress to come to grips with the risk inherent in nuclear technology (as in all human activities), but it did not.[12]

3. "Let the commission decide." Under still another view of the legislative rationale for nuclear regulation, Congress was unwilling to decide either how safe nuclear technology would have to be or even whether there should be a nuclear industry. It therefore delegated the problem to a commission presumably having special expertise in the area. In this view, Congress did not naively think of safety as having an "all-or-nothing" quality, but recognized that there are degrees of safety. Legislators in certain instances went beyond assurances that firms would do their best to prevent accidents to the recognition that there are "varying degrees of bestness."[13]

Even this view, that Congress wanted to have the NRC take over a difficult question on its behalf, can be disputed. In appropriation legislation for fiscal year 1981, the Senate directed the NRC to deter-

10. For analyses of product safety and safety in the workplace, see Walter Y. Oi, "The Economics of Product Safety," *Bell Journal of Economics and Management Science*, vol. 4, no. 1 (Spring 1973), pp. 3-28; W. Kip Viscusi, "The Impact of Occupational Safety and Health Regulation," *Bell Journal of Economics*, vol. 10, no. 1 (Spring 1979), pp. 117-40.

11. Rolph, *Nuclear Power and the Public Safety*, p. 108.

12. The NRC implicitly recognized the absence of a threshold in its 1975 ruling, "Radioactive Material in Light-Water Cooled Nuclear Power Reactor Effluents," Opinion of the Commission, Docket No. RM-50-2, NRC-CLI-75-5, April 30, 1975.

13. See, for example, U.S. Congress, Joint Committee on Atomic Energy, *Governmental Indemnity and Reactor Safety*, Hearings before the Joint Committee on Atomic Energy (85th Cong., 1st sess., 1957), p. 96.

mine and announce an explicit safety goal to guide future regulatory decisions. The House, however, did not join in the request, and the NRC had to continue proceedings on an explicit safety goal without clear congressional backing.[14]

Congress probably never had a consensus that would have led to clear guidance on safety levels and an open rationale for nuclear regulation. Different legislators had different reasons for favoring the regulatory apparatus and policy that emerged. Even today, it would be difficult for Congress or for individual congressmen to embrace publicly any safety standard short of the unattainable zero risk. This is especially unfortunate in the current setting, when intelligent public discussion on the risks of various policies toward nuclear power could lead to much greater understanding.

The Economics of Nuclear Safety

Though Congress did not present a clear or explicit rationale for regulation to enforce nuclear safety, a sound analytical rationale lies in a consideration of the benefits and costs of increasing or decreasing the level of nuclear safety. Before consideration of these benefits and costs, though, it is natural to inquire whether the output of a nuclear plant is needed. Certainly if the electricity from a proposed nuclear plant were not needed, it would make little sense to consider how safe the unneeded plant should be. In fact, the National Environmental Policy Act (NEPA) requires that the need for power from a proposed nuclear facility be established, and the NRC considers such need during its licensing process.[15]

It is difficult to separate the need-for-power issue from safety questions, and at some level of policy making, the issues have to be considered together. The connection between these issues is shown in figure 1. As economists have observed for so long, people find that their "need" for an item decreases as the price increases. The need for power therefore depends on the price that will be charged for the power. Since nuclear safety expenditures become part of the price of nuclear-generated electricity, the level of safety helps determine the price and so the need for power, as indicated by the broken arrow in the figure.

14. The Senate request was contained in the authorization bill, S. 2358 (96th Cong., 2d sess., 1980). See U.S. Nuclear Regulatory Commission, *1980 Annual Report* (March 1981), p. 2.

15. To fulfill this requirement, the NRC has devoted considerable resources to forecasting electricity demands. See NRC, *1980 Annual Report*, pp. 71-72.

FIGURE 1
The Connection between Need-for-Power and Nuclear Safety Issues

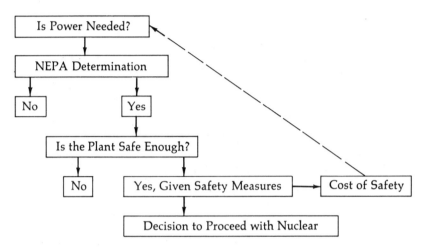

Source: Author.

Before the decision to install nuclear capacity is made, the appropriate authority should have satisfied itself that, at an acceptable level of nuclear safety and safety expenditure, the power is needed by consumers willing to pay a price including the given level of safety expenditure. All too often, a presumed need for nuclear capacity has been the starting point of discussion rather than the result of a decision-making process that involved careful weighing of the alternatives for meeting a perceived need for power.

Assuming for the moment that the output of the nuclear plant is needed and that nuclear power is the least expensive option, let us consider how much nuclear safety should be produced and how much actually would be produced. The costs of guarding against a particular type of accident with known technology have a familiar form. At first, some relatively easy steps can be taken to reduce the chance of an accident or to lessen its consequences. As an accident is made less and less likely, however, it becomes more and more costly to reduce the remaining probability of an accident. As the consequences of an accident are made smaller, it becomes harder to reduce those consequences to zero. The cost of providing additional safety with known technology, therefore, is rather modest at low safety levels, but becomes prohibitive if an attempt is made to wipe out the last increment of known remaining risk.

FIGURE 2
Costs of Producing Safety with Known Technology

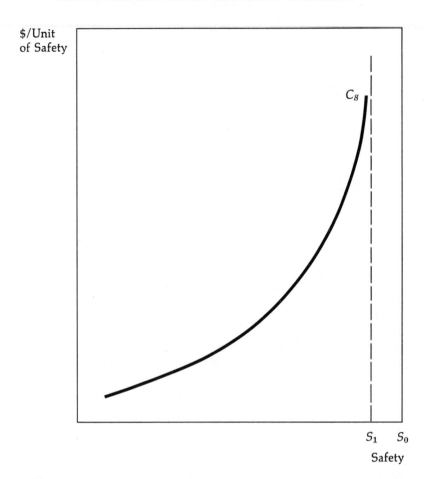

$/Unit
of Safety

C_s

S_1 S_0

Safety

Figure 2 shows the usual pattern of safety costs. With a measure of safety plotted horizontally and the cost of additional safety plotted vertically, the curve rises slowly at first and then more rapidly. The incremental costs of producing safety, C_s, are low at low levels of safety. As some level (S_1) short of complete safety or zero risk (S_0) is approached, the cost of additional safety approaches infinity—indicating that no amount of spending can dispel risk entirely. If society spent all its resources on nuclear safety, the best it could do would be to approach S_1, the greatest degree of safety possible with known technology.

If, however, research into new safety technology is successful,

FIGURE 3
Costs of Producing Safety with Improving Technology

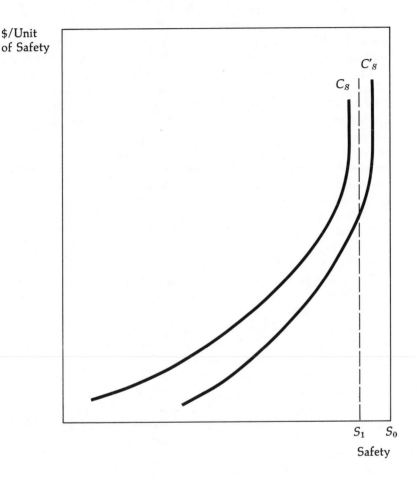

or new information otherwise becomes known, formerly unachievable levels of safety become feasible. Also, each level of additional safety can be achieved at lower cost. Thus it is useful to think of there being at least two safety cost curves. The curve C_s in figure 3 represents the higher cost of safety with currently known technology. The curve C_s' represents the lower cost that would exist with better information and more technically feasible safety measures.

The benefit of reducing the probability and the consequences of a class of accidents is the damage averted, where damage is defined broadly to include pain and suffering as well as the measurable property damage and medical expense. At a low level of nuclear

11

safety, there would be great benefits from the first few safety improvements made. Successive safety improvements would yield smaller additional benefits, until the benefit of removing the last bit of risk from a system already extremely safe would be quite modest. An efficient nuclear safety policy—one conforming to individuals' preferences—would call for the adoption of known safety systems with broadly defined benefits greater than the additional cost of the safety systems. An efficient nuclear safety policy would also call for research into improved safety technologies for as long as the benefit of improved safety technology is greater than the additional cost of the research. At the very least, a rational nuclear safety policy would call for adopting cost-effective technology first and only later pushing to require expensive technology with smaller safety improvements.

Such are the outlines of an economically efficient nuclear safety policy. The safety measures chosen by private firms in the absence of regulation might be different from those that would be efficient. If profit guides the firm's decisions, it will consider only the costs and benefits accruing to it from safety actions. The benefit from improved safety will then be the additional revenue that results from avoiding accidents and thereby avoiding the losses and damage claims for accidents. If there are costs to society that are not costs to the nuclear firm, then the profit-maximizing firm will produce less safety using known technology than would be appropriate. In addition, the firm would have a weakened incentive to do research on reducing the chances and consequences of accidents.

Figure 4 shows the rationale for nuclear safety regulation. Faced with a cost of safety, C_s, and benefits for avoiding accidents, B_f, the profit-maximizing private firm will produce a level of safety S_f. The benefits to the firm from avoiding accidents may not be as large as the total benefits to society from avoiding accidents, B_t. Thus the firm's preferred level of safety, S_f, may be less than society's preferred level of safety based on total benefits, S_t.

Here would lie an airtight economic rationale for some form of nuclear regulation: that the private nuclear industry would not have sufficient incentive on its own to produce the socially desirable level of safety. As described later in this chapter, the private nuclear industry's incentives to avoid serious accidents are deficient under current statute and policy.

These cost-benefit considerations raise three important questions.

1. Is the existing level of safety appropriate? This would be answered, at least in principle, by determining whether there are safety measures with benefits greater than costs that have not been adopted.

12

FIGURE 4
THE RATIONALE FOR NUCLEAR SAFETY REGULATION

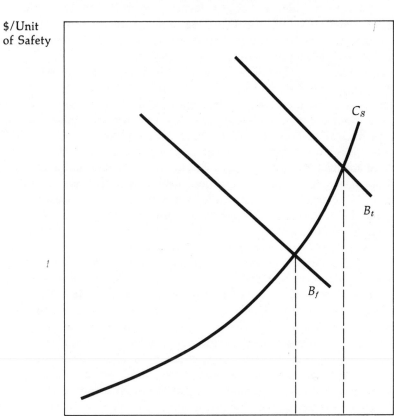

2. Whether appropriate or not, is the existing level of safety being produced cost effectively? Such a question would be answered by exploring whether there are reasons to believe that the current level of safety could be produced at a lower cost, or that the current level of safety spending could yield greater safety.

3. Are sufficient resources being devoted to research that lowers the cost of producing safety? This question could, in principle, be answered by examination of the perceived benefits and costs of safety research.

The consequences of policies leading to too little nuclear safety could be severe. Lives and health might be lost when society was willing and able to prevent the loss. The consequences of policies leading to too much nuclear safety, paradoxically, could be equally severe, if not as visible. If society devotes too much of its scarce resources to nuclear safety, less would be spent on nonnuclear safety. The number of lives lost in nonnuclear areas might be greater than the number of lives saved through nuclear safety.

Some areas of nuclear safety may already have been pursued so far that they are draining precious resources from other, more productive ways of increasing safety. Bernard L. Cohen has calculated, for example, that a standard limiting exposure to radioactive iodine in nuclear reactor effluents is 27,000 times less cost-effective than other ways of saving lives, such as practicing greater safety in the use of medical X-ray procedures.[16] The unmistakable implication is that some people are now losing their lives and health who could be spared if society's resources were reallocated. No one decision maker may be in charge of safety spending on such diverse activities, but the issue nonetheless must be addressed.

In the face of such costs for incorrect decisions, policy makers might understandably consider the course of total indecision on whether to pursue nuclear power as an energy source. Total indecision has its costs, too. When nuclear generation is delayed and fossil fuels are burned to meet the demand for energy, the environmental consequences can be severe. Such consequences can range from increased fatalities from air pollution to the possibility of globally catastrophic changes in climate.[17] Indecision on nuclear power could even undercut the development of alternative energy sources. As long as there is the possibility of a massive expansion of cheap energy from nuclear power, uncertainties will continue to slow down investment in alternative energy sources. It may, therefore, be as important to make some decision on nuclear power as it is to make correct decisions on nuclear safety.

Private Incentives for Safety

As noted earlier in this chapter, the decisions of private firms on nuclear safety will be inappropriate if those firms face deficient incen-

16. Bernard L. Cohen, "Society's Valuation of Life Saving in Radiation Protection and Other Contexts," *Health Physics*, vol. 38, no. 1 (January 1980), p. 36.

17. For an interpretation of the risk of climate change within the context of risk management, see Advisory Committee on Reactor Safeguards (ACRS), "An Approach to Quantitative Safety Goals for Nuclear Power Plants," NUREG-0739 (Washington, D.C.: U.S. Nuclear Regulatory Commission, October 1980), p. 13.

tives to be safe. It might appear that nuclear firms—especially the utilities operating nuclear plants—have ample incentive to be safe. Previous studies have identified at least five strong safety incentives for private firms:[18] (1) the financial loss of the plant, which would exceed the property insurance available on the plant, (2) the loss of revenue from the plant's output, (3) impairment of the utility's ability to provide its service, (4) overall deterioration of the utility's financial picture, possibly leading to bankruptcy, and (5) adverse effects on the entire nuclear industry.

Nuclear firms surely are not cavalier about accidents, in light of these incentives. In the years after the Three Mile Island accident, the plant owner's repeated brushes with bankruptcy served as reminders of the severe consequences of accidents.

Although nuclear firms may have the incentives to produce a high level of safety, their incentives are deficient for providing the extraordinarily high degree of safety that is appropriate. Reasons for the deficient incentives include the firms' strictly limited liability in the event of an accident, their corporate form of organization, and insufficient concern for risk to future generations.

Statutorily Limited Liability. The nuclear industry is unique in the degree to which it is sheltered from damage claims in the event of a serious accident.[19] Airlines and aircraft manufacturers face potentially great liability in the event of a crash; nuclear utilities and especially reactor manufacturers face a much smaller liability in relation to the size of potential damage claims. In addition, nuclear firms are assured that their liability will not exceed a firm ceiling, and part of the insurance to cover the remaining liability has long been provided by the government at token fees.

Liability arrangements for nuclear power date back to the Price-Anderson Act of 1957. This act, amending the Atomic Energy Act of 1954, limited the total liability for a serious nuclear accident to $560 million and provided government coverage for $500 million of the total.[20] Private insurers pooled their resources to offer the other

18. See Legislative Drafting Research Fund of Columbia University, "Issues of Financial Protection in Nuclear Activities," Laurie Rockett, Project Director (December 21, 1973), pp. 4-1 and 4-2. This study is reprinted in U.S. Congress, Joint Committee on Atomic Energy, *Selected Materials on Atomic Energy Indemnity and Insurance Legislation*, pp. 59-197 (Joint Committee Print, 93d Cong., 2d sess., 1974). Also reprinted in this volume is a study by the staff of the Atomic Energy Commission making parallel arguments (p. 20).

19. Dan R. Anderson, "Limits on Liability: The Price-Anderson Act versus Other Laws," *Journal of Risk and Insurance*, vol. 45, no. 4 (December 1978), pp. 651-74.

20. P.L. 85-256, 71 Stat. 576.

$60 million and have increased the coverage offered in more recent years to $160 million, roughly keeping pace with inflation. Under 1975 amendments to the Price-Anderson Act, nuclear utilities have taken over a growing part of the remaining government coverage through contingent obligations to pay $5 million each per reactor in the event of a serious accident. Overall liability, however, remains at $560 million. With decades of inflation, the real level of liability has obviously shrunk. Current law does not call for any upward indexing, though as each new reactor comes into service the pool of contingent obligations will increase by $5 million.[21]

The Price-Anderson Act affects the incentives not only of the utility operating the reactor, but of all other nuclear firms as well. In fixing a single source of $560 million, it relieves reactor manufacturers, architect-engineers, and component suppliers of liability for offsite damages. Thus the powerful influence of potential liability claims on safety is removed from a large part of the industry, leaving only the other private incentives and NRC programs to enforce safety.

The Price-Anderson Act was based on the premise that private insurers were being too cautious in offering insurance for the new technology because they overestimated the risk. This market failure in insurance was to be corrected by government coverage and limited liability. Today there is a genuine question whether private insurers were being too cautious or whether they were correct all along about the size of the risk.[22]

The experience with Price-Anderson again illustrates the connection between safety issues, costs of power, and the need for power. In adopting the Price-Anderson Act, Congress was taking a need for nuclear power development as the starting point. A need for nuclear power should have been the conclusion of a careful consideration of alternatives, with the cost of all alternatives reckoned at the true social cost, not the subsidized apparent cost. The particular insurance subsidy granted by the Price-Anderson Act can contribute to making nuclear power seem inexpensive, just as subsidies to other forms of energy can make them seem inexpensive. But while Price-Anderson can shift risks, it cannot ultimately forestall society's bearing the risks of whatever nuclear power plants are operated.

21. U.S. Nuclear Regulatory Commission, "Financial Protection Requirements and Indemnity Agreements," *Federal Register*, vol. 41, no. 183 (September 20, 1976), pp. 40511-21.
22. William C. Wood, "Nuclear Liability after Three Mile Island," *Journal of Risk and Insurance*, vol. 48, no. 3 (September 1981), pp. 450-64.

Inherently Limited Liability of Nuclear Firms. The Price-Anderson Act may be the most important cause of the inadequacy of private incentives for nuclear safety, but deficient incentives derive from other sources as well. Incentives would remain deficient in the absence of Price-Anderson, if for no other reason than the fact that nuclear firms are corporations. Under the existing legal concept of the corporation, the corporation can be liable for no more than it is worth. If a firm's liability exceeded its worth, it would become bankrupt.

According to the technical literature, the damage from a serious nuclear accident could mount to billions or hundreds of billions of dollars, amounts that easily exceed the worth of a nuclear firm.[23] For such serious accidents, there is no way that the nuclear firm could lose as much as the public. Therefore, the incentive for a firm to guard against such an accident is not as great as society's incentive to prevent the accident.

Risk to Future Generations. Part of the risk of nuclear power is that radioactive waste may enter the environment in the future. A severe nuclear accident is believed to have taken place at a waste disposal facility in the Soviet Union.[24] Since wastes remain radioactive for generations, future generations will bear some of the risk.

The nuclear firm of today—and, indeed, the government of today—has little incentive to consider risks far into the future, since the valuation of the firm or the fate of the government depends little on such risks. The risk to future generations is external to the decision maker of today, so incentives may not be sufficient to generate the appropriate decisions without enlightened and forward-looking regulation.

---

23. A set of estimates can be found in U.S. Nuclear Regulatory Commission, *Reactor Safety Study: An Assessment of Accident Risks in U.S. Commercial Nuclear Power Plants*, NUREG-75/014 (Washington, D.C.: U.S. Nuclear Regulatory Commission, October 1975). This study indicates possible damages of $14 billion at a composite site, but damages could be much higher at heavily populated sites—perhaps as high as $314 billion. See U.S. Nuclear Regulatory Commission, *Technical Guidance for Siting Criteria Development*, NUREG/CR-2239 (Washington, D.C.: U.S. Nuclear Regulatory Commission, November 1982), Appendix C; U.S. Congress, House, Committee on Interior and Insular Affairs, Subcommittee on Oversight and Investigations, "Calculation of Reactor Accident Consequences (CRAC2) for U.S. Nuclear Power Plants (Health Effects and Costs) Conditional on an 'SST1' Release," 97th Cong., 2d sess., November 1, 1982.

24. Richard Corrigan, "Nuclear Disaster—Could Whatever Happened There Happen Here?" *National Journal*, vol. 10, no. 33 (August 19, 1978), pp. 1323-29.

Do Incentives Matter?

We have seen that nuclear firms face some private incentives to produce safety, but that existing law and other factors make incentives for safety deficient. It is important here to ask whether the incentives facing a nuclear firm matter. Does the nuclear firm respond to incentives to produce less safety, and can a firm, in fact, produce less safety? The answer from the industry and regulators alike in the legislative history has been no. In responding, for example, to a suggestion that an uninsured liability of $25 million be imposed on nuclear utilities to correct safety incentives, a government study reported:

> But more fundamentally, it is difficult to accept the proposition that because of nuclear liability insurance, indemnity and limitation of liability, a licensee consciously or unconsciously makes certain decisions pertaining to the health and safety of the general public that it would not make if some portion, say $25 million, of its own assets were exposed to pay public liability claims. Such a view imputes a judgment on the ethics and morality of nuclear licensees that is very questionable.[25]

More recently, at least up until the Three Mile Island accident, the evidence has strongly suggested that nuclear licensees can and do respond to incentives. The following list of findings indicates that incentives matter. The point is not to criticize nuclear firms for not doing everything possible to improve safety, but only to show that the level of safety can be adjusted at the margin and will be if decision makers are motivated to do so.

• Safety at the Three Mile Island plant could have been improved by systematic examination of problems at similar reactors, as described in NRC reports. But plant management concluded that "a formally organized program to prereview and filter the incoming information and subsequently forward it to the appropriate parties would consume more manpower than would be cost effective."[26] This conclusion was set forth in a memorandum dated June 15, 1978—some nine months before the accident.

• In the early minutes of the Three Mile Island accident, plant

25. See the AEC staff study in *Selected Materials on Atomic Energy Indemnity and Insurance Legislation*, p. 20, for this quotation and its context.

26. Memorandum quoted in Stanley M. Gorinson, chief counsel, "Report of the Office of Chief Counsel on the Nuclear Regulatory Commission," submitted to the President's Commission on the Accident at Three Mile Island (October 1979), p. 116. Cited hereafter as "The NRC."

operators were unaware of the potential seriousness of the problem and were concerned with how soon they could bring the plant back on line.[27]

• The operators at Three Mile Island turned off emergency systems providing extra water to the reactor in the mistaken belief that they might be providing so much water that the reactor would be damaged by the pressure. The damaging overpressure would not be an immediate danger, but would require an analysis. "That analysis of vessel mechanics would require taking the plant off line for a matter of weeks or months, with a resulting loss of revenue," testified an engineer for the reactor vendor.[28]

• The supplier of the Three Mile Island reactor felt that it was not responsible for procedures in effect at power stations with its reactors. Nor did it cross-check training procedures on its simulators with the procedures employed later by the trainees. Still, the reactor supplier was the ultimate authority on the design and operations of the reactor.[29]

• An earlier accident, which began in a manner similar to the Three Mile Island accident, generated the information necessary to have prevented the Three Mile Island accident. An investigative body concluded: "Yet through neglect and bureaucratic mistakes that information was never conveyed to the customers [of Babcock & Wilcox, the reactor supplier] prior to TMI-2."[30]

• An investigative panel concluded that the products of a reactor supplier had been combined with too many different designs for the "balance-of-plant," constructed by the architect-engineer, for which the reactor supplier had little or no responsibility.[31]

• A deposition to the staff of the President's Commission on Three Mile Island indicated that nuclear firms are reluctant to report that safety-related modifications are necessary because the licensees may be forced to pay for modifications.[32]

• A chairman of the NRC has said that reactor suppliers are reluctant to propose a safety modification to a plant for fear that the NRC will require that it be supplied to all similar plants.[33]

27. Reactor operator quoted in Stanley M. Gorinson, chief counsel, "Report of the Office of Chief Counsel on the Role of the Managing Utility and Its Suppliers," submitted to the President's Commission on the Accident at Three Mile Island (October 1979), p. 173. Cited hereafter as "The Managing Utility."
28. "The Managing Utility," p. 93.
29. Ibid., p. 107.
30. Ibid., p. 137.
31. Ibid., p. 142.
32. "The NRC," p. 101.
33. Ibid.

- A member of the Advisory Committee on Reactor Safeguards has said that low-level employees of a nuclear utility are not likely to report safety problems requiring large expenditures to correct. "A finding made by an individual deep in an organization which implies heavy costs which is not a regulatory requirement is not likely to be encouraged by what I call the shell of middle management."[34]
- A utility executive openly acknowledged the cost differences between the safest design and the optimum design: "You can design something that is adequate, or you can design something that is 'hell for stout' . . . if you design something that is 'hell for stout,' you are wasting your money."[35]
- A control room design engineer explained why modern human-factors research had not been incorporated to make nuclear plant control rooms less conducive to human error: "The utility industry is by nature a very conservative industry, particularly where operation of the plant is concerned . . . with no motivation to change it and a risk involved in changing it, they tended to stay with it."[36]
- The Advisory Committee on Reactor Safeguards recognized the importance of incentives: "The present system also puts grave responsibility on licensees to make certain that the nuclear technology is used in a way which minimizes the potential for harm to the public even though they have counteractive pressures to minimize costs and improve profitability."[37]
- Concerned about the severe economic penalties associated with locating reactors in remote areas, the nuclear industry has built reactors in highly populated areas.[38]

Again, this list is intended only to suggest that nuclear firms do respond to incentives. Most items on the list concern safety mistakes that were made at Three Mile Island, and steps are being taken to see that these mistakes are not made again.[39] But it would be dangerous to presume that simply avoiding the mistakes brought to light

34. Ibid., p. 102.

35. "The Managing Utility," p. 29.

36. Ibid., p. 33.

37. ACRS, "A Review of NRC Regulatory Processes and Functions," NUREG-0642 (Washington, D.C.: U.S. Nuclear Regulatory Commission, January 1980), p. 1-1.

38. See U.S. Nuclear Regulatory Commission, "Demographic Statistics Pertaining to Nuclear Power Reactor Sites," NUREG-0348 (Washington, D.C.: U.S. Nuclear Regulatory Commission, December 1977).

39. The NRC's detailed operating plan to correct matters following the accident is "NRC Action Plan Developed as a Result of the TMI-2 Accident," NUREG-0660 (Washington, D.C.: U.S. Nuclear Regulatory Commission, May 1980), in two volumes.

by Three Mile Island will be sufficient. There are, no doubt, unknown safety problems that will emerge in the future. Whether these problems are handled better than those of Three Mile Island will depend on whether nuclear institutions truly have been reformed. Much has been done by government and industry since Three Mile Island, but have the key deficiencies been addressed?

The key deficiency pointed out by the Three Mile Island accident, and the deficiency that no amount of technical modification can correct, is the dispersion of knowledge on safety and of the incentive for safety. Nuclear safety is primarily a problem in the use of knowledge. Investigative bodies inquiring into the institutional difficulties that led to the Three Mile Island accident found a fragmentation of knowledge and the responsibility for using it. It is this fragmentation—and not technical or even plant operation problems—that poses the true underlying threat to nuclear safety.

The nuclear utility is held responsible for the safe operation of its plants. Yet the utility faces sharply limited financial responsibility for the consequences of an accident. The reactor supplier is the ultimate safety authority on its reactors, but faces no liability for offsite damages. The architect-engineer designs and builds the plant complex, but has virtually no responsibility for its safety once it is turned over to the utility. The NRC has incentives to promote safety and the appearance of safety, but lacks the intimate knowledge of the plant that the licensee has. Thus, no one party has both an adequate incentive to promote safety and the knowledge of how to do it.[40]

The Legislation and Today's NRC

The economic rationale for nuclear safety regulation is that because of deficient private incentives, private firms will not produce an efficient level of safety. The NRC's legislative mandate—to the limited extent that it has a coherent mandate—is not well suited to regulation aimed at overcoming deficient safety incentives.

The NRC can choose from two types of policies to promote safety. First, it could seek out the sources of deficient safety incentives and the areas where they apply, then attempt to strengthen incentives in those targeted areas. Second, it could simply mandate that particular safety systems and procedures be installed, and then use inspection and enforcement tools to see that its mandates are carried out. The second course requires that the NRC know a great

40. For evidence supporting a similar conclusion by the staff of the President's Commission on the Accident at Three Mile Island, see "The Managing Utility," pp. 44-50, 61, 66-74, 150.

deal about how safety can best be produced. This direct regulation is consistent with the NRC's legislative guidance in a way that incentive strategies are not. After all, Congress passed and twice renewed the Price-Anderson Act, openly removing a major incentive to safety and counting on direct regulation to take up the slack.

If the NRC's mandate is to make nuclear technology "safe" by finding thresholds of exposure or accident probability below which no harm will be done, then the NRC cannot fulfill its mandate. Such technically determinable thresholds would make life easier for policy makers because the technology would then be simply "safe" or "unsafe." But such thresholds have been extremely elusive.

If the NRC's mandate is to take a sticky decision off the hands of the legislative branch and to make hard decisions about the role of nuclear power in society, then it is not well constituted to carry out its mandate. Over half of the agency's staff consists of scientists and engineers, and the agency is geared toward making technical decisions, not social ones. The NRC has a great deal of expertise in the risks of nuclear power and how to control them. It has no particular expertise in making broad social judgments on the acceptability of those risks. As Harold Green put it in his often-quoted statement, "No elite group of experts, no matter how broadly constituted, has the ability to make an objective and valid determination with respect to what benefits people want and what risks people are willing to assume in order to have these benefits."[41]

The NRC is set up as a technical decision-making body. With a vague mandate telling it only "to assure public health and safety," it is finding itself called upon to make social decisions for which it is not well equipped. Fundamental improvements in nuclear regulation will have to await some resolution of the NRC's role.

41. Harold Green, "Risk-Benefit Calculus in Safety Determinations," *George Washington University Law Review*, vol. 43 (1975), p. 792, quoted in Rolph, *Nuclear Power and the Public Safety*, p. 101.

2
The NRC and Its Safety Programs

When the NRC was formed in 1975, the idea of having a separate regulatory body in charge of nuclear safety was not new. Within the old Atomic Energy Commission, the regulators had long been organized in a regulatory division separate from the promoters. An agency entirely for the regulators had been proposed as early as 1956.[1] By 1961, a study team at the University of Michigan Law School actively supported a complete separation of regulation from the AEC. At that time the congressional Joint Committee on Atomic Energy (JCAE) favored an independent licensing board within the AEC. The AEC, however, prevailed in its plan for keeping the regulators independent but within the agency and reporting to the commission.[2]

An important factor in keeping regulatory responsibilities within the AEC for so long was the fear that expertise in nuclear technology was insufficient to staff two separate agencies. This fear was openly expressed by JCAE members in 1956 and was resolved to the satisfaction of some in Congress only in 1973–1974 when the bill splitting off the NRC was being considered.[3]

The Arab oil embargo in 1973–1974 led to strong support for federal programs to develop new energy sources. The AEC, with its expertise on nuclear energy, became part of the agency created to explore new energy resources, the Energy Research and Development Administration (ERDA). The regulators of nuclear safety did not fit well into ERDA, so the time was ripe for a separate nuclear safety agency. The proposed formation of ERDA, combined with a perceived cover-up of nuclear safety problems by the AEC, was sufficient legislative incentive for establishing a Nuclear Energy Commission. (The name was changed to the Nuclear Regulatory Commission during the legislative process to emphasize its nonpromotional nature.)

1. Elizabeth S. Rolph, *Nuclear Power and the Public Safety: A Study in Regulation* (Lexington, Mass.: Lexington Books, 1979), pp. 40-41.

2. Ibid., p. 48.

3. See the remarks of Representative Frank Horton of New York in the *Congressional Record*, vol. 119, part 33 (December 19, 1973), p. 42573.

Allegations of a cover-up of nuclear safety problems by the AEC came to a head in 1973 when a suit filed under the Freedom of Information Act forced disclosure of an updated study of the possible consequences of a nuclear accident. Instead of the possible 3,400 deaths, 43,000 injuries, and $7 billion in property damage estimated in 1957, a working group found that a nuclear accident could cause 45,000 deaths, 70,000–100,000 injuries, and $17 billion in property damage.[4] The forced release of these results added to fears that, in its promotion of nuclear power, the AEC was subordinating safety. The splitting off of the NRC has, therefore, sometimes been interpreted as a punitive measure against the AEC.[5]

It is not clear what Congress expected from the new NRC. One supporter of the reorganization expressed hope that faster licensing of nuclear plants would occur, with "due consideration" to safety.[6] Another wanted the reorganization to prevent crucial safety issues from being buried at low levels in the bureaucracy, so that they would be aired and resolved by the commissioners.[7]

Whatever Congress expected, it is clear in hindsight that the creation of the NRC brought no fundamental changes in nuclear regulation at first. "The creation of the NRC did not materially change nuclear power plant licensing," observed the Advisory Committee on Reactor Safeguards. The NRC, staffed mostly by the AEC regulatory bureaucracy, adopted AEC regulations and therefore the AEC approach to safety.[8] Perhaps more important, the fundamental problems of technical uncertainties about nuclear power and lack of congressional guidance on the trade-off between energy and risk remained unresolved. Almost five years after the NRC's establishment, an NRC commissioner was able to testify, "I still think it is fundamentally geared to trying to nurture a growing industry."[9]

It is difficult to characterize the attitude toward safety that has

4. The working papers of this study, which was never formally published, are in the "WASH-740 Update File" at the U.S. Nuclear Regulatory Commission Public Document Room, 1717 H St., N.W., Washington, D.C. 20555.

5. Rolph, *Nuclear Power and the Public Safety*, p. 116.

6. Remarks of Representative William S. Moorhead of Pennsylvania, *Congressional Record*, vol. 119, part 33 (December 19, 1973), p. 42589.

7. Remarks of Senator Abraham Ribicoff, *Congressional Record*, vol. 120, part 21 (August 15, 1974), p. 28606.

8. Advisory Committee on Reactor Safeguards (ACRS), "A Review of NRC Regulatory Processes and Functions," p. 3-2. See also Stanley M. Gorinson, chief counsel, "Report of the Office of Chief Counsel on the Nuclear Regulatory Commission," submitted to the President's Commission on the Accident at Three Mile Island (October 1979), p. 14, cited hereafter as "The NRC"; Rolph, *Nuclear Power and the Public Safety*, p. 155.

9. Quoted in "The NRC," p. 2.

prevailed at the NRC since its formation in 1975. At least two important pressures on NRC commissioners and staff, however, can be identified. The first is the pressure to appear to be reducing the risk of harm from the use of nuclear technology. "Looking safe" is especially important in the absence of clear and objective measures of the actual safety of plants. The safety philosophy of the NRC has been described as "basically a risk reduction philosophy without fully articulated safety goals."[10] This approach has not yet specified how safe is safe enough, nor does it quantify the amount of risk implicitly being accepted by society if delay of nuclear generation leads to the use of alternative and risky technologies.

Aside from the pressure to appear to be reducing safety, there is the pressure not to regulate the nuclear industry out of existence. With their substantial specialized knowledge of nuclear processes, many NRC employees would certainly be worse off if the nuclear industry ceased to exist. Within the AEC, these two pressures combined to make the AEC's early safety goal the adoption of the strictest requirements consistent with the commercial viability of nuclear power.[11] A similar criterion would be a natural result of the pressures bearing on the NRC. It must be seen as promoting safety, but a literal policy of safety at all costs would find the NRC presiding over the dismantling of its regulated industry.

Organization of the NRC

Little of the nuts-and-bolts work of the NRC is done by the commissioners or by offices reporting directly to the commission. Most of such work is done in the three offices set up by the law establishing the NRC and in other program offices set up by the commission early in the NRC's life. The commissioners have a relatively minor role in the licensing of individual facilities; the adjudicatory hearings are conducted by Atomic Safety and Licensing Boards that nominally report to the commission. Even the first level of appeal of licensing decisions is handled by Atomic Safety and Licensing Appeal Boards and is out of the commission's hands.

The NRC was loosely organized and lacked coordination, at least up until the Three Mile Island accident. A senior NRC official has said that the program offices operate as independent "fiefdoms."[12] In addition to the program offices, the NRC comprises the commission

10. ACRS, "An Approach to Quantitative Safety Goals for Nuclear Power Plants," NUREG-0739 (Washington, D.C.: U.S. Nuclear Regulatory Commission, October 1980), p. 51.

11. Rolph, *Nuclear Power and the Public Safety*, p. 77.

12. "The NRC," p. 24.

staff (six offices), the support staff (nine offices), and the adjudicatory boards and panels that actually hear licensing matters.

With such a diversity of offices, it would be important for the NRC to have a well-coordinated executive authority. For reasons of inertia as well as of history, however, the NRC has long had two focuses of executive authority, and at times has seemed to lack coherent leadership. The internal separation of regulatory authority at the AEC was only the beginning of a trend that isolated commissioners more and more from the actual business of regulation, especially licensing. A strict ex parte rule, carried over from the AEC days, prevents commissioners' involvement in licensing matters that may come before the commission in its adjudicatory role. As a result, the commissioners have little knowledge of the day-to-day conduct of regulation.[13] Since the 1960s at the AEC, the commission and staff have been housed in separate locations, and today the commission and staff are separated by the distance from downtown Washington to the Maryland suburbs.

As a result, competing centers of influence—the commission and the executive head of the staff—came into being. By 1976, the JCAE staff had recognized that the commissioner serving as NRC chairman could not administer day-to-day activities. "Even if he did, he is much too removed and isolated from the day-to-day problems by the layer upon layer of management in the organizational structure."[14] Adding further to the lack of coordination was the absence of authority for the staff's chief executive, the executive director for operations. Program offices could either take their concerns directly to the commission or go through the executive director for operations, the nominal chief executive.

Figure 5 shows the organization of the NRC after the reforms —effective October 1, 1980—that followed the Three Mile Island accident. The chairman and the staff under the executive director for operations remain organizationally as well as physically distant. The four program offices and the nine support offices—shown in the lower half of the chart and located in the Maryland suburbs of Washington—are responsible for most of the NRC's day-to-day regulation. They report to the executive director for operations, who in turn reports to the chairman of the commission. The chairman and the commissioners, located in downtown Washington, D.C., have authority over the offices shown in the upper half of the chart—the Office of Public Affairs, the Office of Congressional Affairs, the com-

13. Ibid., p. 29.
14. Ibid., p. 14.

FIGURE 5

Organizational Chart of the U.S. Nuclear Regulatory Commission

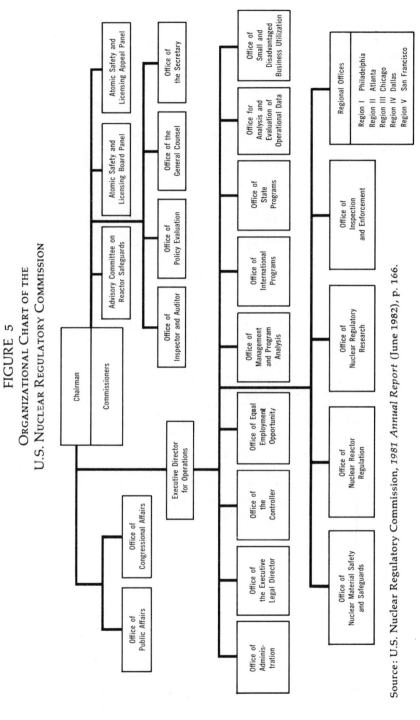

Source: U.S. Nuclear Regulatory Commission, *1981 Annual Report* (June 1982), p. 166.

mission's own staff, and the boards and panels responsible to the commission as a whole. The reforms following Three Mile Island strengthened the executive roles of the chairman and the executive director for operations, but did not resolve the presence of two separate staffs—one far from the reach of the chairman.

A further difficulty is that the role of the commission itself is not well defined. As discussed earlier, the commission can be thought of primarily as a social or a technical decision-making body, with little congressional guidance to define its role. The commissioners' professional backgrounds have ranged from law to nuclear physics. It is also not clear what role the commissioners are intended to play in the management of a nuclear emergency. Before the Three Mile Island accident, in fact, it was not expected that the commissioners would play any role during an accident.[15] NRC personnel can take control of a plant in an emergency, but these personnel would presumably be technical professionals from the staff.[16]

Types of Regulation

In effect, NRC regulations arise in three main ways: rule making, Regulatory Guide issuances, and licensing proceedings.

Rule Making. The NRC, like other regulatory agencies, has broad authority to adopt rules governing the behavior of licensees. Through the publication-and-comment procedure, proposed rules make their way through the *Federal Register* and into Title 10, chapter 1 of the *Code of Federal Regulations*. Because of their broad applicability, rule makings offer the possibility of resolving safety issues just once and then having the decision apply to all licensees. Matters such as reactor siting and planning for emergencies show promise for a one-time treatment in rule making, and both siting and emergency planning have been the subjects of recent NRC rule making. The disadvantage of rule making lies, at least in part, in its broad applicability. In important technical ways, each nuclear power plant is different, so there are good reasons for deciding many matters on a case-by-case basis.

Regulatory Guides. Regulatory Guides describe methods acceptable to the NRC staff for complying with specific parts of NRC regulations. They are not legally binding. The added cost of demonstrating that an alternative complies with regulations, however, encourages

15. Ibid., p. 33.
16. Ibid., p. 133.

28

compliance with Regulatory Guides. Regulatory Guides have been described as "excessively prescriptive," setting out detailed ways of meeting regulations and not leaving sufficient leeway for licensees to take initiative to come up with new methods of promoting safety.[17]

Licensing Proceedings. An important part of nuclear regulation has been developed through the licensing proceedings required before any nuclear facility can be constructed and then operated. It is, therefore, worth a careful look to understand how the licensing process operates.

Licensing is conducted in two stages. In the first stage, the utility seeks permission to build the plant. In the second stage, conducted as the plant nears completion, the utility seeks a license to operate the plant.

To begin the process of getting a construction permit, the applicant files information on the proposed plant's safety and effect on the environment. The application generally consists of ten or more large volumes. Public notice of the application is given and the NRC review begins. The NRC staff reviews safety aspects, effect on the environment, security for the nuclear materials, and antitrust aspects of the proposed plant. Independently, the Advisory Committee on Reactor Safeguards (ACRS)—a panel of distinguished scientists serving part time—conducts its own review.

The application for a construction permit is subjected to a mandatory public hearing before a three-member Atomic Safety and Licensing Board. The board is chaired by a lawyer; the other two members are an environmental scientist and a technical expert (nuclear engineer or reactor physicist). The licensing board's decision whether to issue a construction permit can be appealed to an Atomic Safety and Licensing Appeal Board, then to the commission, and finally to the federal court system.

Site preparation and advance work on the plant can begin before the construction permit is approved under the terms of a limited work authorization from the NRC. The utility can save as much as seven months by starting work before the construction permit is issued, but it has to submit its environmental data early to get an authorization.

Nuclear plants in the United States have in recent times taken ten years or more to build and license for operation. The operating license review begins two to three years before the scheduled com-

17. The view that Regulatory Guides have been excessively prescriptive is stated by the ACRS, in "A Review of NRC Regulatory Processes and Functions," p. 4-2.

pletion of the plant and proceeds in much the same manner as the construction permit proceeding. The applicant files for an operating license, and the NRC staff and Advisory Committee on Reactor Safeguards review the application. At this stage, a public hearing is optional. The same possibilities for appeal are open, including recourse to the federal courts. At the end of the process, a license is issued with whatever restrictions and conditions are felt necessary for the plant to operate safely. Throughout the licensing process there is a strong presumption that nuclear power technology can be made acceptably safe; the NRC has never denied an operating license to a constructed nuclear facility.[18]

To the proponents of nuclear power, the licensing process is too slow, riddled with redundancy, too uncertain, and too susceptible to delay from the maneuverings of environmental groups. To the opponents of nuclear power, the licensing process is an elaborate charade with a predetermined outcome and little opportunity for meaningful public participation. Much of the conflict over licensing results from ambiguity of just what licensing should be. Some of the more important issues of the conflict are:

• Social or technical? As in other nuclear safety decisions, the issue arises whether licensing ought to consider overall social goals or simply narrow technical matters concerning the plant at hand.

• Adversary or collegial? There is some question whether licensing ought to be a judicial "fight" between the applicant and the intervenors (and perhaps the NRC staff) to bring out the truth, or a cooperative technical venture to meld the opinions of involved professionals.

• Open or private? Should the process be entirely open, or should there be room for behind-the-scenes negotiation and compromise?

• "Meet the rules" or go beyond? Should licensees be required only to meet the relevant regulations, or should they be prodded—by the Advisory Committee on Reactor Safeguards, for example—to advance safety beyond that of the last plant approved?

• Once-for-all decisions or not? There is some sentiment for making key decisions—such as the basic acceptability of a site—early in the process and making it very hard to reverse the decision later.

18. This is not to say that expensive safety modifications will not be ordered after the operating license is issued. Indeed, in the case of the Indian Point 1 reactor, first licensed in 1962, emergency core-cooling requirements led the utility to close the plant after it had gone on line. For a review of the emergency core-cooling controversy, see Irvin C. Bupp and Jean-Claude Derian, *Light Water: How the Nuclear Dream Dissolved* (New York: Basic Books, 1978), pp. 133-34. For a description of nuclear licensing, see NRC, *1981 Annual Report* (June 1982), p. 6. See also "The NRC," p. 45.

Such an approach would avoid the costs of indecision, but could lead to decisions that would later be regretted.

• ACRS as adviser or participant? On the one hand, the Advisory Committee on Reactor Safeguards, composed as it is of eminent scientists, could bring substantial expertise to licensing. On the other hand, a greater role in licensing and heavier membership responsibilities could sap the very expertise that is the strength of the ACRS.

• When to backfit? How important must a safety improvement be to justify changing designs before a plant is completed ("ratcheting") or requiring modifications on completed plants ("backfitting")? The answer depends crucially on whether current reactors are viewed primarily as embodying experimental technology or mature technology. If the technology is experimental, then licensing is only authorization for the experiment to proceed at a specific site. In this view, the regulator and licensee work together to be sure the results of the experiment are incorporated into plant operations. There is no presumption that the initial licensing sets the final conditions under which the plant may operate. But if the technology is viewed as mature, licensing is authorization to go ahead with a commercial process under specified conditions. Only substantial new evidence or discovery of errors in licensing would justify backfitting.

The American system of nuclear licensing decides social and technical issues in an open, adversarial atmosphere, with few final decisions and with design changes always a possibility. By contrast, most European countries have narrowly technical licensing conducted out of public view in a spirit of compromise. Outside of Germany, Europeans opposed to nuclear power wage their battles in the political arena rather than at specific plant licensing actions.[19] The European model may offer some important lessons for nuclear regulation in the United States, particularly in the separation of technical from social issues.

Enforcing Nuclear Regulation

Enforcing the NRC's safety regulations is the job of the Office of Inspection and Enforcement. This office had a staff of 975 in fiscal year 1981, with a fifth of the staff at headquarters in Bethesda, Maryland, and four-fifths assigned to regional offices in Philadelphia, Atlanta, Chicago, Dallas, and San Francisco.

The Inspection and Enforcement staff engages both in routine inspections and in "reactive" inspections, made in response to reports

19. Michael W. Golay, "How Prometheus Came to Be Bound," *Technology Review*, vol. 82, no. 7 (June/July 1980), pp. 29-30.

of safety problems from sources ranging from the licensee to members of the public. At the end of fiscal year 1981, the NRC had 124 inspectors assigned to specific plant sites; these resident inspectors provide day-to-day contact between licensees and the NRC.[20]

Following the Three Mile Island accident, the NRC has adopted a more vigorous inspection policy with more emphasis on direct inspection of problems. Earlier policy had amounted to inspecting to validate the licensee's own internal inspection process. The NRC's new enforcement policy calls for classifying violations into six levels of severity, and then applying progressively more stringent measures for more serious, continued, or repeated violations. Licensees who identify and solve safety problems on their own, or show good faith, are to face reduced sanctions. The object of the NRC policy is to assure in the long term "that noncompliance is more expensive than compliance."[21]

Three principal enforcement actions are available to the NRC. One action is a notice of violation, a written notice describing a violation of a legally binding requirement. The licensee is required to pledge, usually under oath, to take corrective action. A civil penalty, a second possible action, is a monetary penalty for violating important rules or reporting requirements. The maximum civil penalty for a violation was raised from $5,000 to $100,000 in 1980, and a provision limiting total civil penalties in any thirty-day period to $25,000 was eliminated. Since each day of noncompliance could be treated as a separate violation, continuing violations can become quite expensive for the licensee.[22] Another enforcement action is an order, a written NRC directive. An order can require modifications to a license or action as severe as immediate shutdown.

Actions in addition to these three include enforcement conferences with the licensee, a variety of bulletins and notices, Notices of Deviation from nonbinding safety practices, and Immediate Action Letters confirming a licensee's agreement to take safety action. The NRC has broad authority to order whatever mix of enforcement actions it finds appropriate.

An important issue in enforcement is whether licensees are or should be relied on to report safety data and safety problems. Independent investigators have several times accused the NRC of over-

20. For a more detailed account of the resident inspector program, see NRC, *1981 Annual Report*, pp. 87-88.

21. NRC, "Proposed General Statement of Policy and Procedure for Enforcement Actions," *Federal Register*, vol. 45, no. 196 (October 7, 1980), pp. 66754-61.

22. Ibid., p. 66754.

reliance on data generated by the nuclear industry.[23] The problem apparently is not a lack of reports; the NRC receives ten to fifteen Licensee Event Reports daily. These reports are often incomplete and difficult to interpret, apparently written up just to satisfy statutory reporting requirements. The flood of reports, in the NRC's own words, "had overwhelmed the capacity of the NRC or the industry to assimilate the lessons of experience."[24]

A simple solution to incomplete reporting of nuclear safety data would be to penalize under-reporting more heavily or to involve the NRC more closely in plant operation. Such solutions, though, might not work and could make matters worse. Heavy penalties for self-reported violations reduce the incentive to report. As the Advisory Committee on Reactor Safeguards put it, "Although it is difficult to excuse mistakes and unintended violations of regulations, the threat of legal jeopardy in such instances can only create an environment of protective cover-up among the threatened that tends to hide important safety information."[25] As for greater NRC involvement in plant operation, even starting up and shutting down reactors, the committee stated, "Thus far, the NRC has avoided this because it would essentially relieve the licensees of any responsibility for design and operational decisions."[26]

The NRC has taken at least one step designed in part to provide data independent from that reported by the nuclear industry—dosimeters to measure radiation levels have been installed around reactor sites. According to the NRC the dosimeters cover all sectors of the compass, population centers and "high public interest locations" extending approximately ten miles from the plants.[27]

Research and Other Safety Programs

Research on Nuclear Safety. The Office of Nuclear Regulatory Research is the NRC's most heavily funded office, receiving just over half of the annual budget. The research office, however, employs only 10 percent of the NRC's personnel—indicating that most NRC research is done outside the agency by independent contractors. NRC employees coordinate research and direct the program.

23. See "The NRC," p. 96, and references cited there.
24. NRC, *1980 Annual Report*, p. 91.
25. ACRS, "A Review of NRC Regulatory Processes and Functions," p. 7-10, fn. 7.5.
26. Ibid., pp. 7-4–7-5.
27. NRC, *1980 Annual Report*, p. 143.

In its early years, the agency emphasized "confirmatory research." Nuclear regulations had been made, the agency thought, using pessimistic or conservative values for key parameters at every stage. Regulations were, therefore, more stringent than they had to be. Confirmatory research was aimed at discovering just how conservative the regulations were, to provide confidence that safety was being ensured, and perhaps to allow some relaxation of requirements found entirely too conservative.

Following the Three Mile Island accident, new research demands have been placed on the NRC. The accident drew attention to potential accident sequences that had not been dealt with adequately, and much effort was redirected to those accident sequences. The NRC was reluctant to give up major parts of its confirmatory research programs, but, as the Advisory Committee on Reactor Safeguards put it, "The NRC may have to reduce sharply some research which is merely confirmatory in nature where there is good reason to believe that the current regulatory requirements provide adequate protection to the public."[28]

Some NRC research is done in facilities that duplicate or simulate actual conditions of a commercial nuclear reactor. The Loss-of-Fluid-Test and Semiscale programs are examples. In other cases, the processes being researched are so far beyond those that can be physically simulated or statistically managed that theoretical simulations are necessary. It would not be sensible, for example, to melt down tons of reactor fuel to see what would happen in a full-scale meltdown at a power reactor. It would also not be practical to wait years for a component to fail just to be sure it had a very low probability of failure. In such cases a computer code or a theoretical model is used. Much of the NRC's research effort is geared toward assessing and verifying these codes and models. In addition to dealing with power reactors and current technology, the NRC does research on the nuclear fuel cycle and environmental effects, as well as research on the advanced reactor technologies that are still being developed.

Fuel Cycle, Waste Management, and Other Programs. Rounding out the NRC's safety programs are programs that are indirectly related, if at all, to nuclear power reactors. The NRC Office of Nuclear Material Safety and Safeguards is responsible for regulating nuclear materials. This responsibility includes the reactor fuel cycle from the processing of uranium to the handling of spent fuel. Since nuclear

28. ACRS, "Comments on the NRC Safety Research Program Budget for Fiscal Year 1982," NUREG-0699 (Washington, D.C.: U.S. Nuclear Regulatory Commission, July 1980), p. 3.

materials are commonly transported—sometimes long distances—at various stages of the fuel cycle, transportation safety is a significant responsibility for the Office of Nuclear Material Safety and Safeguards. Other nuclear materials under NRC jurisdiction include a wide range of industrial and medical substances.

A key issue in the long-term future of nuclear power is how highly radioactive wastes of nuclear power plants can be disposed of. These wastes will retain radioactivity for thousands of years. The NRC has a hearing in progress to determine what degree of confidence there is that wastes can be safely disposed of. The NRC shares responsibility for waste management with the Environmental Protection Agency and the Department of Energy. The NRC's role in waste management can be expected to be quite limited until the Department of Energy has decided on a waste storage technology consistent with EPA standards.[29]

The NRC conducts a variety of nonregulatory activities, as the diversity of offices on the agency's organizational chart suggests. It has the usual offices of congressional and public affairs, as well as others, ranging to an office of equal employment opportunity and an office of small and disadvantaged business utilization.

Special Factors at the NRC

Having surveyed the organization of the NRC, we can now turn to special factors, some inherent in the agency's mission, that make nuclear regulation a difficult task.

Vast Uncertainties about the Nuclear Risk. Every human undertaking faces some uncertainties, but those about nuclear regulation are so vast they deserve special attention. As it turns out, no one is certain about even the most basic parameters defining the nuclear risk. There is uncertainty over the likelihood of nuclear accidents, and there is uncertainty over the consequences. There is substantial uncertainty even about the effects of the small amounts of radiation routinely released from operating reactors. In nuclear safety issues, the uncertainty is often expressed in orders of magnitude—powers of ten, such as one thousand or ten thousand. A weather forecast is a sure bet by comparison.

No good estimate exists, therefore, of just how severe the consequences of a serious nuclear accident might be, though at different times and under varying assumptions the damages have been esti-

29. On this point, see ACRS, "A Review of NRC Regulatory Processes and Functions," p. 4-9.

mated at $7 billion, $17 billion, $14 billion—and, more recently, $314 billion. There is also no good estimate of the probability of a serious nuclear accident. The one study that makes a serious attempt to estimate these probabilities indicates that an accident may be only a fifth as likely or as much as five times as likely as its central estimates. Peer review has established, however, that even this wide range of probabilities is too narrow to take in the range of uncertainty.[30]

Technical Dissent. In the face of such uncertainty, it is natural that parties in the licensing process and even within the NRC will have differing opinions. A challenge to nuclear regulators is to resolve these differences into clear and consistent policy. Because the range of uncertainty is so wide, however, reasonable people may take greatly divergent yet equally plausible positions on matters of nuclear safety.

The NRC has not always handled technical dissent well. Early in the agency's life, a staff member and three vendor engineers resigned because of safety concerns they felt were being hidden. Following the resignations and a congressional investigation, the NRC adopted policies aimed at promoting a full and free expression of technical dissent even when it conflicted with the judgment of supervisors.[31] These policies had not fully succeeded at the time of the Three Mile Island accident; an NRC inspector had been trying for months to get the agency to recognize the very safety problems that allowed the accident to happen.[32] After the accident, the NRC renewed its efforts to prevent suppression of technical dissent.

The NRC has to walk a very fine line on technical dissent. If, on the one hand, genuine professional differences about safety are suppressed, vital safety information can be lost. Further, the ultimate release of information about safety measures left untaken and safety

30. The early studies are: "Theoretical Possibilities and Consequences of Major Accidents at Large Nuclear Power Plants," WASH-740 (Washington, D.C.: U.S. Atomic Energy Commission, March 1957); and "WASH-740 Update File," NRC Public Document Room. The probabilistic study is *Reactor Safety Study: An Assessment of Accident Risks in U.S. Commercial Nuclear Power Plants,* NUREG-75/014 (Washington, D.C.: U.S. Nuclear Regulatory Commission, October 1975). The peer review referred to is that of Harold W. Lewis, chairman, "Risk Assessment Review Group Report to the U.S. Nuclear Regulatory Commission," NUREG/CR-0400 (Washington, D.C.: U.S. Nuclear Regulatory Commission, September 1978).

31. For background on the resignations and the NRC response, see U.S. Congress, Joint Committee on Atomic Energy, *Investigation of Charges Relating to Nuclear Reactor Safety,* Hearings before the Joint Committee on Atomic Energy, vols. 1-2 (94th Cong., 2d sess., 1976).

32. "The NRC," pp. 108, 112.

concerns left unresolved can make the agency look bad. On the other hand, if technical dissent is resolved simply by adopting the most conservative position proposed by any staff member, the result is a dictatorship in safety standards by the most risk-averse person. When one person dictates safety standards, the whole purpose of having teams of professionals is defeated. "Too much safety" in a world of limited resources means giving up other goods and amenities, perhaps even giving up lives and health in pursuing nonnuclear energy.

Unresolved Safety Issues. Given the uncertainty inherent in questions of nuclear safety, it is not surprising that licensing actions generate safety issues that cannot immediately be resolved. Since many safety issues were common to a group of nuclear power plants, they were often labeled "generic." Critics charged that the NRC removed serious safety issues from the licensing process simply by declaring them generic and deferring them for resolution "later." More recently, generic issues have returned to the licensing process with the requirement that no plant be licensed if the license is found inappropriate in light of the unresolved issues. A long and growing backlog of generic issues did, in fact, build up during the late 1970s. Although the NRC has taken steps to cut the backlog of generic issues, it is not yet clear how effective these steps will be.[33]

Difficulties with Modern Management Methods. NRC efforts to use modern tools of agency management have had, at best, only mixed success. Early in the agency's history an attempt was made to use a comprehensive program of management by objectives (MBO). In such a program, the staff would identify those objectives crucial to the agency's mission and—with commission approval—implement plans to achieve the objectives. The MBO program would be a management and information system, with provisions for reporting progress toward attainment of objectives at various "milestones" or key subobjectives.

The MBO program, imposed by support staff onto program offices, was met unenthusiastically. The managers who were supposed to benefit from the MBO program regarded it as just another reporting system. Program offices chose dates and goals that they would have achieved in the absence of the MBO program. Thus the

33. See NRC, "NRC Program for the Resolution of Generic Issues Related to Nuclear Power Plants," NUREG-0410 (Washington, D.C.: U.S. Nuclear Regulatory Commission, January 1978); "Task Action Plans for Unresolved Safety Issues Related to Nuclear Power Plants," NUREG-0649 (Washington, D.C.: U.S. Nuclear Regulatory Commission, February 1980); *1981 Annual Report*, pp. 13-16.

program offices reported the accomplishment of objectives on schedule, and the MBO program was reduced in fact to a reporting program.[34]

A modern management tool with somewhat more success is cost effectiveness analysis, adopted throughout the agency in 1978. The NRC has labeled such analysis "value-impact" analysis, making an analogy with benefit-cost analysis and indicating that benefits and costs include more than strictly monetary benefits and costs. The analogy with benefit-cost analysis is imperfect, however, because benefit-cost analysis implies analysis of different levels of an activity and a choice of the most favorable level. The NRC's value-impact guidelines apparently rule out the possibility of a finding that a lower level of safety could be preferred to a higher level of safety. Thus the analysis concerns alternative ways of reaching a stated safety goal and is properly known as a cost-effectiveness analysis.[35]

The commission is formally committed to value-impact analysis. So far it appears to have fared better than MBO. It is difficult to quantify many of the unknowns required to reach a conclusion through value-impact analysis. Still, if value-impact analysis forces policy makers to consider alternative ways of producing safety and if it generates now rare information on the costs of producing safety, it will prove worthwhile.

34. See NRC, *Annual Report 1975* (1976), for an early statement of objectives to be used in managing the agency. By the time of the *Annual Report 1978* (1979), the word "objectives" had been dropped from the description of the mission of the responsible NRC office.

35. For a general discussion of issues involved in value-impact analysis, see the commissioner action paper, "Value-Impact Guidelines," SECY-79-424, July 2, 1979, available at the NRC Public Document Room.

3
Results of Nuclear Safety Regulation

Nuclear regulation has increased the amount of resources going to nuclear safety. As this chapter will show, however, those resources have not been used cost effectively. It is not possible, therefore, to say with any certainty that nuclear regulation has increased the level of safety very much over what it would be in the absence of regulation. In fact, because so much uncertainty exists over measurement of nuclear safety and the associated costs and benefits, we may never know whether the level of safety being achieved is the appropriate one. This chapter examines the reasons for this pessimism about our knowledge of nuclear safety. Uncertainty about nuclear power, however, does not automatically justify a decision to forgo nuclear power as a source of energy. Nor does uncertainty justify a decision to put off deciding about nuclear power. Since the alternatives to nuclear energy—conservation or other energy sources—involve uncertainties of their own, uncertainty cannot be avoided simply by halting nuclear development.

This chapter evaluates the evidence on the results of nuclear regulation. Then an effort will be made to identify the causes of these results. The discussion addresses five questions: (1) How safe are nuclear power plants, and how credible is the evidence on their safety? (2) Is the current level of safety at least minimally acceptable? (3) Can the current level of safety be justified on cost-benefit grounds? (4) Is the current level of safety being produced cost effectively? (5) What accounts for the current state of nuclear safety?

Definitive answers are probably even harder to find in nuclear regulation than in other forms of regulation. Nonetheless, it is useful to identify what is known about the state of nuclear safety.

How Safe Are Nuclear Power Plants?

The earliest studies of the consequences of nuclear accidents had little to say that was of direct usefulness to the policy maker. The first study, "Theoretical Possibilities and Consequences of Major

Accidents at Large Nuclear Power Plants," was commissioned in the 1950s when Congress was considering government insurance for nuclear plants and wanted to know how large the loss might be.[1] Its often-cited figures of 3,400 deaths, 43,000 injuries, and $7 billion in property damage are essentially a worst-case estimate. This study therefore told policy makers of the 1950s that a nuclear accident might be quite severe. Nuclear policy makers could find this only slightly more useful than aviation policy makers would find a conclusion about the consequences of a loaded airliner crashing into the Super Bowl. It suggests that consequences might be catastrophic, but gives little idea of the chances of such an accident or how to defend against the more likely accidents of regulatory interest.

When the initial study was revised in 1964–1965 to account for the larger reactors then being designed, the revised figures were again of little direct relevance. The new figures indicated that there could be as many as 45,000 deaths, 70,000–100,000 injuries, and as much as $17 billion in property damage.[2] These figures told policy makers little more than that the worst possible accident with the larger reactors would be much more severe than the worst possible accident with the earlier reactors. The main result of the revised report was probably to accelerate the loss of confidence in the AEC eight years later when the results were finally made public under pressure of a suit filed under the Freedom of Information Act.

Policy makers needed a study that (1) estimated the probability and consequences of the whole spectrum of accidents, not just the worst, and (2) estimated the costs and benefits of changes in the required level of safety. Though the second task of estimating costs and benefits has never been seriously attempted, there has been an ambitious attempt to accomplish the first, a very difficult and controversial task. The study was the subject of controversy even before it was released in final form in October 1975. The report is known variously as WASH-1400 (its official document number), the *Reactor Safety Study*, and the Rasmussen Report, after project director Norman Rasmussen.[3]

1. "Theoretical Possibilities and Consequences of Major Accidents at Large Nuclear Power Plants," WASH-740 (Washington, D.C.: U.S. Atomic Energy Commission, March 1957). Hereafter cited as WASH-740.

2. "WASH-740 Update File," U.S. Nuclear Regulatory Commission Public Document Room, 1717 H St., N.W., Washington, D.C. 20555.

3. U.S. Nuclear Regulatory Commission, *Reactor Safety Study: An Assessment of Accident Risks in U.S. Commercial Nuclear Power Plants*, NUREG-75/014 (Washington, D.C.: U.S. Nuclear Regulatory Commission, October 1975). Draft version published by U.S. Atomic Energy Commission under document number WASH-1400, August 1974.

The Reactor Safety Study. The *Reactor Safety Study* concluded that the chance of a serious nuclear accident was very small. The study identified a melting of the nuclear core as the main threat to release significant amounts of radioactivity offsite. Its calculations indicated that the chance of a core melt was only one in 5,000 per reactor per year (per "reactor-year"). Moreover, indications were that even if the core melted, radioactivity would most likely be contained. The *Reactor Safety Study* estimated that most core melts would not cause even one fatality at the time of the accident ("early" or "prompt" fatality, as opposed to a delayed or "latent" fatality). But rare extreme core melts, with probabilities ranging downward to one in a billion per reactor-year, could cause as many as 3,300 early fatalities, 45,000 early illnesses, and $14 billion in property damage.[4]

How credible is the *Reactor Safety Study?* This question can be answered by reference to two important reviews of its methods and results. The first, and perhaps the sharpest technical criticism of the study, was issued by the Union of Concerned Scientists (UCS).[5] A somewhat more sympathetic review—but one equally harsh on some specific points—was conducted by an NRC-chartered review group headed by Harold W. Lewis.[6]

The *Reactor Safety Study* made its probability estimates by attempting to consider all events that could initiate a serious accident and then by tracing those events to the possible outcomes. At each step in the "fault tree" or "event tree," probabilities of failure were assigned, based on sources ranging from empirical data to engineering judgment. Experience could be used, for example, to assign a probability to failure of a particular component. Then implications of the component failure could be traced to other events that ultimately would lead to an accident, such as failure of backup components or human error. At each branch of the tree, a probability would be assigned so that an overall probability for the sequence of events could be derived.[7]

This procedure is subject to serious difficulties, as the UCS critique points out.[8] It is impossible to be sure that all potential

4. Ibid., p. 83.

5. Henry W. Kendall, study director, *The Risks of Nuclear Power Reactors: A Review of the NRC Reactor Safety Study* (Cambridge, Mass.: Union of Concerned Scientists, 1977). Hereafter cited as *Risks of Nuclear Power Reactors.*

6. Harold W. Lewis, chairman, *Risk Assessment Review Group Report to the U.S. Nuclear Regulatory Commission,* NUREG/CR-0400 (Washington, D.C.: U.S. Nuclear Regulatory Commission, September 1978). Hereafter cited as *Risk Assessment Review Group Report.*

7. For a discussion of methodology, see *Reactor Safety Study,* chap. 4.

8. *Risks of Nuclear Power Reactors,* p. 132.

accident sequences have been considered. Any omission would bias the final probability estimate downward. In fact, the *Reactor Safety Study* explicitly excluded some potential causes of accidents, such as component aging, serious earthquakes, sabotage, and terrorism. UCS also found the elementary data for estimating component reliability to be incomplete or uncertain.[9]

The *Reactor Safety Study* frequently considered the probability of a safety system working as designed. The UCS saw this as a serious shortcoming, involving as it did the implicit assumption that the safety system working as designed would halt the accident. UCS pointed out that a safety system might be correctly designed to perform an incorrect function. Such a safety system could work perfectly and still not end the accident. Another objection by UCS was that the *Reactor Safety Study* had used misleading summary statistics from some of the probability distributions.[10]

Still another serious difficulty with the *Reactor Safety Study*'s approach was incomplete treatment of failures that would knock out several components at the same time. These common-mode failures make the basic fault tree or event tree invalid, since events occurring along the branches of the tree are not statistically independent as assumed.[11]

The Lewis committee reviewed technical criticisms and found merit in many of them. It was still able to conclude that the *Reactor Safety Study* provides "the most complete single picture of accident probabilities associated with nuclear reactors."[12] The committee said that fault-tree and event-tree analysis with an adequate data base is the "best available tool" for quantifying accident probabilities. The committee recognized, however, that it is impossible to be sure all possible accident sequences have been included. It therefore recognized that any such calculation is subject to doubt about its completeness. The committee said it was unconvinced that a proper treatment had been given to the possible roles of fires, earthquakes, and human error.[13]

The Lewis committee agreed that the data on component reliability were inadequate and that the *Reactor Safety Study* had not fully accounted for common-mode failures. The committee was especially

9. Ibid., chaps. 2, 5.
10. Ibid., pp. 20-21; chap. 4.
11. Ibid., chap. 3.
12. *Risk Assessment Review Group Report*, p. viii.
13. Ibid., pp. 14-15, 24-26, 27-28.

critical of the study's use of square-root bounding of some parameters —taking arbitrary high and low values, then multiplying them together and taking the square root for a central estimate.[14]

As for the effect of the errors on the overall accident probabilities in the *Reactor Safety Study*, the Lewis committee said it could not determine whether the probabilities were overstated or understated. And despite the great range of admitted uncertainty in the *Reactor Safety Study*, the Lewis committee said the actual uncertainties were even greater than admitted.[15]

UCS had sharp criticism also for the nonprobabilistic part of the *Reactor Safety Study*—the calculation of consequences of various radioactive releases. The UCS critique charged that the *Reactor Safety Study* had been too optimistic on the rate of radiation dispersion, the effectiveness of evacuation, and the incidence of biological effects. UCS also criticized summary tables of accident consequences as misleading in understating the number of latent deaths that would come years after a nuclear accident.[16]

The Lewis committee agreed that improvements were needed in the biological effects models. It also acknowledged that some latent effects of radiation were not prominently displayed in the study. It said, however, that these effects were calculated systematically in study appendixes for the first time.

Table 1 lists *Reactor Safety Study* results, as well as additional results for accident effects calculated in technical appendixes not prominently displayed in the main report. Table 2 lists alternative results based on what UCS calls "correction of a number of the most obvious *RSS* errors."[17] The UCS calculations assume probabilities twenty times higher, early consequences ten times higher, and latent consequences several times higher. Neither set of results contains the consequences of the worst conceivable accident, since consequences were calculated for a representative reactor site instead of the most unfavorable site.

The Sandia Study. Just how severe an accident at an unfavorable site might be was dramatized in November 1982 with the release of some site-specific calculations performed by Sandia National Laboratories

14. Ibid., pp. 6-9.
15. Ibid., p. viii.
16. *Risks of Nuclear Power Reactors*, chaps. 6, 7, 8; pp. 118-25.
17. Ibid., p. 113.

TABLE 1
ACCIDENT CONSEQUENCES CALCULATED IN THE REACTOR SAFETY STUDY

Chance per Reactor-Year	Early Fatalities	Early Illnesses	Property Damage, 10^9	Latent Cancer Fatalities[a]	Thyroid Nodules[a]	Genetic Effects[a]	Total Latent Fatalities[b]	Money Damages,[c] 10^9
2×10^{-4}	<1.0	<1.0	<0.1	<30	<30	<30	80[d]	0.012[e]
1×10^{-5}	<1.0[f]	<1.0[f]	0.15[f]	810	6,000[f]	40[f]	1,100	0.319
1×10^{-6}	<1.0	300	0.9	5,100	42,000	750	6,700	1.937
1×10^{-7}	110	3,000	3	13,800	105,000	1,800	17,900	5.799
1×10^{-8}	900	14,000	8	25,800	180,000	3,300	32,900	13.351
1×10^{-9}	3,300	45,000	14	45,000	240,000	5,100	54,800	23.424

a. These figures are presented per year in the *Reactor Safety Study*; totals are obtained by multiplying by thirty years. Thus the figures above represent effects summed over the thirty-year period after the accident in which the latent consequences would manifest themselves. See *Reactor Safety Study*, p. 83.

b. Total latent fatalities were calculated but not presented in the *Reactor Safety Study*'s main report. These figures, based on *Reactor Safety Study* results, are presented in *Risks of Nuclear Power Reactors*, p. 123.

c. Based on monetary equivalents presented in Tobias W. T. Burnett, "The Human Cost of Regulatory Delays," *Nuclear Technology*, vol. 33, no. 2 (mid-April 1977), pp. 203-11.

d. Interpolated (on logs) from *Risks of Nuclear Power Reactors*, p. 122, figure 10-3.

e. Latent fatalities only, because of lack of point estimates in other categories.

f. Read from *Reactor Safety Study*, figures 5-3, p. 88; 5-4, p. 89; 5-6, p. 91; 5-7, p. 92; and 5-8, p. 93.

SOURCE: *Reactor Safety Study: An Assessment of Accident Risks in U.S. Commercial Nuclear Power Plants*, NUREG-75/014 (Washington, D.C.: U.S. Nuclear Regulatory Commission, October 1975), p. 83, tables 5-4 and 5-5, with corrections and adjustments as noted.

TABLE 2

Accident Consequences Calculated in Risks of Nuclear Power Reactors

Chance per Reactor-Year[a]	Early Fatalities[b]	Early Illnesses[b]	Property Damage,[c] 10^9	Latent Cancer Fatalities	Thyroid Nodules[d]	Genetic Effects[d]	Total Latent Fatalities	Money Damages,[e] 10^9
2×10^{-4}	<10	<10	0.15	810	36,000	240	5,864	1.057
2×10^{-5}	<10	3,000	0.9	5,100	252,000	4,500	36,390	6.550
2×10^{-6}	1,100	30,000	3	13,800	630,000	10,800	97,200	18.405
2×10^{-7}	9,000	140,000	8	25,800	1,080,000	19,800	179,700	38.570
2×10^{-8}	33,000	450,000	14	45,000	1,440,000	30,600	303,900	70.903

a. Presented here per 1 reactor-year (original expressed per 100 reactor-years). Each figure is twenty times the corresponding probability from the *Reactor Safety Study*.

b. These are ten times the corresponding *Reactor Safety Study* results; see *Risks of Nuclear Power Reactors*, p. 119: "Correction of Prompt Risk of Nuclear Accidents."

c. Property damage figures are said to be understated, but no numerical correction is offered in *Risks of Nuclear Power Reactors* (see p. 127).

d. These figures are six times the corresponding *Reactor Safety Study* results, incorporating a doubling of exposure and a tripling for radiobiological response. See *Risks of Nuclear Power Reactors*, p. 123.

e. Based on monetary equivalents presented in Burnett, "The Human Cost of Regulatory Delays."

Source: Henry W. Kendall, study director, *The Risks of Nuclear Power Reactors: A Review of the NRC Reactor Safety Study* (Cambridge, Mass.: Union of Concerned Scientists, 1977), p. 125, table 10-2, with additions and adjustments as noted.

for the NRC.[18] As compiled and reported by a House subcommittee staff, the calculations showed that a severe accident at the Salem reactor in New Jersey could take 100,000 lives; a severe accident at the Indian Point reactor in New York could cause $314 billion in property damage.[19] Because of the widespread publicity given the Sandia calculations, it is worth some effort to understand the nature and the limitations of the calculations.

The Sandia study was not an attempt to redo the entire *Reactor Safety Study*. Its much narrower mandate was to provide guidance for siting reactors by calculating how accident consequences might vary among potential sites. The study did not attempt to determine the probabilities of various accidents as the *Reactor Safety Study* did. Instead it simply assumed a set of accidental releases of radiation and calculated the consequences of such releases. Most press attention focused on "Siting Source Term 1" or "Group 1" releases, the worst ones studied, which would involve a large-scale fuel melt followed by loss of installed safety features and a severe direct breach of containment.[20]

Consequences were calculated by Sandia using an improved version of the *Reactor Safety Study*'s "Calculation of Reactor Accident Consequences" (CRAC) model. Sandia's CRAC2 contains improved modeling of the way that released radiation would interact with weather conditions, better simulation of possible evacuations, and revised risk factors for latent cancer.[21] The Sandia report, as released by the NRC, presents mean or expected average consequences for a stated hypothetical release from ninety-one sites. Clearly the mean consequences, as the more or less typical result of a severe accident, are important to consider. A severe accident could combine, however, with unusually unfavorable offsite conditions to make consequences far more severe than the mean. The "worst cases" calcu-

18. U.S. Nuclear Regulatory Commission, *Technical Guidance for Siting Criteria Development*, NUREG/CR-2239 (Washington, D.C.: U.S. Nuclear Regulatory Commission, November 1982); U.S. Nuclear Regulatory Commission, "Estimates of the Financial Consequences of Nuclear Power Reactor Accidents," NUREG/CR-2723 (Washington, D.C.: U.S. Nuclear Regulatory Commission, September 1982).

19. U.S. Congress, House, Committee on Interior and Insular Affairs, Subcommittee on Oversight and Investigations, "Calculation of Reactor Accident Consequences (CRAC2) for U.S. Nuclear Power Plants (Health Effects and Costs) Conditional on an 'SST1' Release" (97th Cong., 2d sess., November 1, 1982), pp. 6, 10.

20. See NRC, "Estimates of the Financial Consequences of Nuclear Power Reactor Accidents," p. 2, for a listing and descriptions of other assumed releases.

21. For a description of CRAC2 and the changes, see NRC, *Technical Guidance for Citing Criteria Development*, Appendix E.

TABLE 3
Accident Consequences for the Indian Point Site

Accident	Early Fatalities	Early Injuries	Cancer Fatalities	Property Damage, 10^9	Money Damages,[a] 10^9
Mean	716	3,136	6,988	10.2	11.4
Worst calculated	50,000	167,000	14,000	314.0	331.2

a. Based on monetary equivalents in Tobias W.T. Burnett, "The Human Cost of Regulatory Delay," *Nuclear Technology*, vol. 33, no. 2 (mid April 1977), p. 205. SOURCE. U.S. Nuclear Regulatory Commission, "Estimates of the Financial Consequences of Nuclear Power Reactor Accidents," NUREG/CR-2723, scaled for Unit 3 by technique on p. 4; U.S. Congress, House, Committee on Interior and Insular Affairs, Subcommittee on Oversight and Investigations, "Calculation of Reactor Accident Consequences (CRAC2) for U.S. Nuclear Power Plants (Health Effects and Costs) Conditional on an 'SST1' Release" (97th Cong., 2d sess., November 1, 1982), p. 6.

lated by Sandia, though, were not released or endorsed by the NRC. These worst cases were obtained and released by a House subcommittee and involved a severe accidental release of radiation carried over population centers by wind, then suddenly deposited in a "rainout." Whether the mean cases or worst cases were of more relevance was a major issue in public statements accompanying release of the results.

What, then, is the citizen or policy maker to make of the Sandia results? Three conclusions are warranted:

1. Consequences of accidents may vary a great deal, depending on the site. For example, the Sandia report indicates that the number of fatalities expected from a given hypothetical release might range from as many as 830 at a densely populated site to as few as 0.07 (most likely no one would be killed) at a favorable site.[22]

2. The uncertainties of what might happen after an accident, even holding the site constant, are vast. As table 3 shows, the mean consequences of an accident at the densely populated Indian Point site include 716 early fatalities and $10.2 billion in property damage. The table also shows, however, that unfavorable offsite conditions could push the number of early fatalities to 50,000 and property damage to $314 billion.

22. See ibid., pp. C-1–C-6.

3. No matter how bad a worst case is conceived, a still worse one can be imagined. It is now clear that the worst accident prominently reported in the *Reactor Safety Study*, involving 3,300 early fatalities and $14 billion in property damage, is far from being truly a worst case. Indeed, no modern-size reactor had worst-calculated damages as low as $14 billion in the House subcommittee's release of the Sandia results. Even the 50,000-fatality, $314 billion damage case for the Indian Point reactor is not the worst imaginable; it is always possible to assume less effective evacuation or even worse weather.[23]

The Sandia study is part of a continuing effort to understand the consequences of accidents. Though the November 1982 releases of information made the publicly available consequence estimates larger, it is possible that future work will show them to have overstated the risk.[24] The Sandia study made no effort to determine accident probabilities and therefore has narrower scope than the *Reactor Safety Study*, which remains the most complete technical attempt to estimate the nuclear risk directly. We now turn to some indirect methods of inference about nuclear safety.

The Denenberg Estimate. An alternative estimate of the nuclear risk has been made by Herbert S. Denenberg, former insurance commissioner of Pennsylvania. Denenberg did not attempt a technical study like the *Reactor Safety Study*, but instead estimated the probability of a serious nuclear accident based on the premiums charged by private insurers of nuclear power plants.[25]

Before examining the Denenberg inferences, it is important to consider whether any meaning can be attached to probabilities inferred from insurance premiums. Since uncertainty about accident consequences is so great, insurance premiums can at best be regarded as an educated guess. Still, the premium is the educated guess of a party —the insurer—that faces rewards for correct judgment and penalties for incorrect judgment. Insurers may be unique in the nuclear industry

23. For a summary of the most important uncertainties in the Sandia study, see ibid., chap. 2.

24. Recall that the Sandia study simply assumed a release of radioactivity of given size. There is some evidence that the assumed release was too large. See ibid., pp. 2-18–2-23.

25. Herbert S. Denenberg, "Testimony before the Atomic Safety and Licensing Board, Licensing Hearing for Three Mile Island Nuclear Power Plant," November 7, 1973, reprinted in U.S. Congress, Joint Committee on Atomic Energy, *Possible Modification or Extension of the Price-Anderson Insurance and Indemnity Act*, Hearings before the Joint Committee on Atomic Energy (94th Cong., 2d sess., 1974), pp. 226-40.

TABLE 4

Accident Consequences Inferred from Insurers' Behavior

Coverage Level 10^6	Premium ($ per $1 million of coverage)	Fraction of Premiums Available for Claims	Expected Value of Claims ($ per $1 million of coverage)	Inferred Change per Reactor-Year
1	32,500	0.58	18,850	1.885×10^{-2}
2–5	16,250	0.58	9,425	9.425×10^{-3}
6–10	6,500	0.58	3,770	3.77×10^{-3}
11–20	3,250	0.58	1,885	1.885×10^{-3}
21–40	1,625	0.58	942.5	9.425×10^{-4}
41–60	1,000	0.58	580	5.8×10^{-4}
61–80	1,000	0.58	580	5.8×10^{-4}
81–95	1,000	0.58	580	5.8×10^{-4}

SOURCE: Calculated by the author from method used in Herbert S. Denenberg, "Testimony before the Atomic Safety and Licensing Board, Licensing Hearing for Three Mile Island Nuclear Power Plant," November 7, 1973, reprinted in U.S. Congress, Joint Committee on Atomic Energy, *Possible Modification or Extension of the Price-Anderson Insurance and Indemnity Act*, Hearings before the Joint Committee on Atomic Energy (94th Cong., 2d sess., 1974), pp. 226–40.

in the extent to which they bear direct financial consequences for incorrect judgment. Thus the implications of their guesses may be more compelling in some ways than those of the *Reactor Safety Study* group, who had little to lose if their estimates were overly optimistic.

In making his inferences, Denenberg noted that insurers charged $32,500 for the first million dollars of coverage on a typical nuclear reactor. The premium fell off for additional coverage levels until $40 million of coverage was reached, then leveled off at $1,000 per million dollars of coverage. With 42 percent of premiums needed to cover expenses other than losses, 58 percent of premiums would be available as a reserve to cover losses. Dividing ($1,000 x 0.58) by $1 million of coverage, Denenberg inferred that the chance of an accident per reactor-year was 0.00058, or about 1 in 1700. This 1973 estimate based on insurers' behavior indicates that accident probabilities are many times higher than indicated by the *Reactor Safety Study*. Table 4 lists probabilities of accidents at each level of coverage, calculated using the Denenberg method. The accident probabilities range from around 1 in 53 (1.885×10^{-2}) for an accident causing only $1 million in offsite damage to 1 in 1700 (5.8×10^{-4}) for an accident causing $40–95 million in offsite damage.

It should be clear that this 1 in 1700 probability is an inferred probability. Insurers did not say that the chance of a serious nuclear accident would be 1 in 1700 per reactor-year. Instead, they accepted a gamble that would pay off only if the probability turned out to be less than 1 in 1700.

The Denenberg estimate contains a number of weaknesses, even if it is considered valid to make inferences about this risk from insurance behavior. First, the inferences apply only up to the limit of coverage offered by private insurers ($95 million at the time Denenberg wrote) and not to the accidents of special policy interest beyond. Second, the inferences take for granted that insurance premiums return only normal profits, and not any profits beyond. Even defining "normal profits," however, depends on firm knowledge of accident probabilities and consequences. It is hard to define what a normal profit would be for taking on a risk as unusual as nuclear power. Compounding the problem is the belief that the probability of a nuclear accident is not only uncertain but also very low. If the insurer charged a premium based on a very small chance of an accident, these premiums would mount into a sizable fund only very slowly. An early accident caused by bad luck—even if the probability of an accident were in fact low—would cause the insurer to suffer a large loss.

The Denenberg estimates, therefore, have their problems just as those of the *Reactor Safety Study* do.

The Naive Actuarial Approach. To rely only on data on the absence of a particular class of accident for a specified number of years in making statistical inferences is naive; such data are usually used to establish an upper limit on the probability of an accident. If, for example, 600 reactor-years have been recorded (a figure reached in 1981) without a catastrophic accident, the chance of a catastrophic accident per reactor-year is taken as being no greater than one in 600.

This figure, however, is not an absolute upper limit. There is about a 50–50 chance that 600 reactor-years would pass without an accident even if the chance of an accident were as high as 1 in 400.[26] Moreover, accidents may become more likely as facilities grow older and components age. It is true that the naive actuarial approach is quite pessimistic. The evidence of 600 reactor-years of operation

26. Suppose that the rate of occurrence of accidents characterizes an exponential distribution and that the chance of an accident is 1/400 per reactor-year. Then the chance of having an accident in 600 years is $F(T) = 1 - \exp(-\mu T) = 1 - \exp(-1/400 \times 600) = 0.51$. The chance of not having an accident would be 0.49. A society could get lucky and not have an accident just as a gambler could get lucky and flip ten consecutive heads with a coin. Or the society might have an early accident, just as the first toss of a coin could come up tails.

without a catastrophic accident would involve, by itself, a probability of 0/600, or zero. Inferring a probability of 1/600 assumes, contrary to fact, that a serious accident did occur—or will occur in the next instant. Naive actuarial estimates can be used as plausible upper-range estimates, but they do not establish an absolute upper limit.

Evidence from Reactor Operating Experience. The evidence on nuclear safety from reactor operating experience can be interpreted in strikingly contrasting ways. Nuclear proponents point to the excellent statistical safety record in the industry. A favorite statistic is that no member of the public has been killed in a radiological accident at a licensed commercial nuclear power plant. The statistic is correct, though nonmembers of the public (employees) have been killed in nonradiological accidents at commercial reactors and in radiological accidents at noncommercial reactors. Further, though the increase in cancer from radiological effluents of operating plants is not large enough to be statistically distinguishable, it is entirely possible that some member of the public has died from cancer induced by radiation from a nuclear plant. Nonetheless, the statistic of zero early fatalities among members of the public is an enviable one, and one not matched by other industries.

Nuclear opponents point to radiation exposures among nuclear power plant workers and to "near misses" involving safety to dispute the claim of a safe operating record. There have been some significant near misses, including a fire that knocked out key safety systems and the temporary failure of one reactor to shut down upon command.[27]

The one event that has come to dominate discussions of nuclear safety is the March 1979 accident at Three Mile Island. Measured in dollars, the Three Mile Island accident was not severe, at least to members of the public. Public damage claims were settled for $25 million—an amount far below the loss of the operating utility. Measured by public health effects, the accident, again, was not severe. The low levels of radiation released are not expected to cause a statistically distinguishable change in cancers among the exposed population.[28]

In spite of these measures, however, Three Mile Island dealt a severe blow to the credibility of optimistic risk estimates like the

27. These near misses and others are documented in a quarterly report series from the NRC to Congress, "Abnormal Occurrences." See U.S. Nuclear Regulatory Commission, *1981 Annual Report* (June 1982), pp. 54-58.

28. Mitchell Rogovin, director, Special Inquiry Group, *Three Mile Island: A Report to the Commissioners and to the Public*, NUREG/CR-1250 (Washington, D.C.: U.S. Nuclear Regulatory Commission, January 1980), p. 153. Hereafter cited as *Report to the Commissioners*.

Reactor Safety Study. An accident causing the damages of Three Mile Island would have been predicted to occur once in 33,000 reactor-years, but actually occurred after 500 reactor-years. Only 1/66 of the expected time to the accident had elapsed before it occurred. If the chance of such an accident really is 1/33,000, then there is only a 1.5 percent chance that the accident would occur so soon.[29] Based only on the Three Mile Island accident—and it is only one event—expected values of damages in the *Reactor Safety Study* were understated by a factor of 66. The ad hoc corrections of the Union of Concerned Scientists, placing damages higher by a factor of 60, are much more consistent with Three Mile Island than are findings in the *Reactor Safety Study.* Even the inferred insurance estimate of the chance of this accident, 1 in 1060 per reactor-year, is much closer to the achieved probability than that in the *Reactor Safety Study.*

The most severe damage that Three Mile Island did to the perception of the safety of nuclear power was not, however, in discrediting probability estimates. The most severe damage may have been done by the public impression that the utility and NRC alike were inept and confused in responding to the accident.[30]

Summary of Risk Estimates. Figure 6 is a graphic summary of the risk estimates reviewed above. The horizontal axis represents dollar values of damages, and the vertical axis marks off the probability that an accident causing given dollar damages will occur. A downward-sloping curve in the figure indicates that the probability of worse and worse accidents becomes lower and lower. A curve near the lower left corner of the diagram represents an optimistic assessment of the risk, with damages and probabilities low. A curve farther to the right reflects a more pessimistic view, with higher probabilities of more damaging accidents. Note that because the axes are scaled logarithmically, moving one mark upward involves a tenfold increase in damages.

The curve closest to the origin in figure 6 represents the spectrum of accidents calculated in the *Reactor Safety Study.* Its damage figures, however, do not include damages for loss of lives and health.

29. The figure of 1/33,000 per reactor-year is drawn from a curve that has been used in the past by NRC as though it showed the probabilities of various levels of legally determined damages. See NRC, "Financial Protection Requirements and Indemnity Agreements," *Federal Register,* vol. 41, no. 183 (September 20, 1976), pp. 40511-21. There is, however, some doubt whether this curve adequately represents the conclusions of the *Reactor Safety Study.* The 1.5 percent chance referred to derives from an exponential distribution with a mean of 1/33,000: $1 - \exp(-\mu T) = 1 - \exp(-500/33,000) = 0.015$.

30. For a critical look, see *Report to the Commissioners,* especially chaps. 3-17.

FIGURE 6
ESTIMATES OF THE NUCLEAR RISK

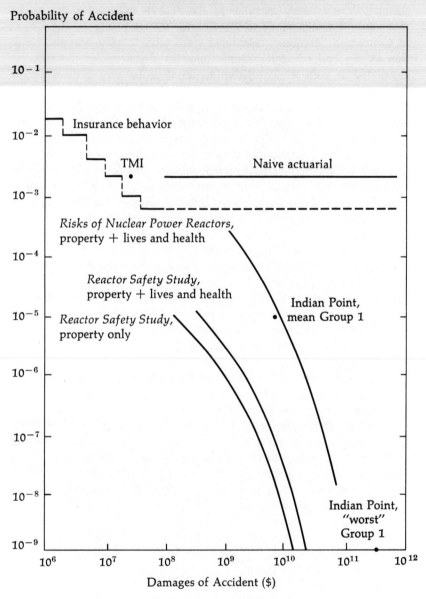

Source: Author's plot of tables 1–4; *Reactor Safety Study*, p. 83.

The next curve out from the origin represents total damages, including compensation for loss of lives and health as calculated from the *Reactor Safety Study* in table 1. The curve third from the origin represents the corrections to the *Reactor Safety Study* by the Union of Concerned Scientists, calculated from table 2. The stepped function at the upper left represents insurers' estimates of the risk, as calculated in table 4. The naive actuarial probability is drawn in at the upper right.

Three individual accidents, one actual and two hypothetical, are indicated by large dots on the diagram. The relative frequency and damages of the Three Mile Island accident are denoted by the point "TMI." A hypothetical accident considered the mean outcome of a Group 1 release at the Indian Point site, as calculated in the Sandia study, is labeled "Indian Point, mean Group 1." This accident clearly involves larger damages than accidents of similar probability in the *Reactor Safety Study*. The worst accident calculated by Sandia and released by the House subcommittee is labeled "Indian Point, 'worst' Group 1." Keeping in mind the logarithmic scaling of the axes, it is clear that this accident would be far worse than the worst accident indicated in the *Reactor Safety Study*.

Two aspects of the diagram are notable. First, it is clear that the various estimates of the nuclear risk are widely divergent. Second, the Three Mile Island accident appears to fit much better with the more pessimistic estimates of nuclear risk.

Is the Level of Nuclear Safety Minimally Acceptable?

Before considering whether the level of nuclear safety is appropriate on cost-benefit grounds, one should consider whether the safety level is at least minimally acceptable. One basis for a consensus answer of yes is the failure of various inquiry groups investigating Three Mile Island to recommend a shutdown of operating plants. Another answer, that nuclear safety must be greater than it was at the time of the Three Mile Island accident, is implicit in the report of the president's commission that studied the accident: "Whether in this particular case we came close to a catastrophic accident or not, this accident was too serious. Accidents as serious as TMI should not be allowed to occur in the future."[31]

The acceptability of the current degree of nuclear safety depends on the breadth of an individual's perception of nuclear power. For

31. John G. Kemeny, chairman, *The Need for Change: The Legacy of TMI*, Report of the President's Commission on the Accident at Three Mile Island (October 1979), p. 15.

those who perceive it as an energy alternative to fossil-fired plants, the present safety level is acceptable since society seems to accept the pollution, disease, and death associated with coal plants in return for coal-generated electricity. The health effects of coal generation can be severe; generation of 10^{10} kilowatt hours using coal is estimated to kill 10 to 200 people.[32]

For those who see nuclear power not just as an energy source but as a symbol of disturbing trends in society, no amount of nuclear safety is enough, and nuclear safety issues are only a vehicle for opposition. In such a view, the disturbing trends symbolized by nuclear plants include most prominently the proliferation of nuclear weapons. Little middle ground exists between this position and the position that nuclear power is an energy source only, and a benign one compared with the alternatives. Since this conflict over nuclear technology is not likely to be resolved before policies must be established, policy makers should recognize that the substantial consensus on nuclear power as an energy source is that it is sufficiently safe not to shut down all operating plants immediately.

Is the Level of Nuclear Safety Efficient?

As discussed in chapter 1, an appropriate, or efficient, level of nuclear safety using known technology will be achieved when the benefits of increasing safety are equal to the costs of increased safety. In terms of figure 4, showing benefits and costs of safety, all technologies with benefits greater than their incremental costs should be adopted. If society goes beyond an efficient level, nuclear safety will be achieved at the expense of goals more worthwhile at the margin than the additional nuclear safety.

To determine the efficient level of nuclear safety, appropriate measures of benefits and costs are necessary. Both benefits and costs are difficult to formulate, let alone estimate.

The benefits of additional nuclear safety consist of the reduced likelihood of nuclear accident damages, with "damages" defined to include pain and suffering as well as conventionally calculated outlays for evacuation and medical care. How much does a given safety system reduce the chance of an accident? The difficulty of answering this question can be seen in the *Reactor Safety Study*, where probabilistic methods like fault-tree analysis had to be relied on. The reduction in accident probability contributed by any safety system is

32. Advisory Committee on Reactor Safeguards (ACRS), "An Approach to Quantitative Safety Goals for Nuclear Power Plants," NUREG-0739 (Washington, D.C.: U.S. Nuclear Regulatory Commission, October 1980), p. 4.

costly and difficult to calculate. How much can safety systems mitigate accident damage, even if they do not make an accident less likely? As the controversy over the nonprobabilistic portion of the *Reactor Safety Study* shows, this too is a difficult technical problem.

Even with perfect technical estimates of the effects of safety systems, however, a benefit measure would be elusive. First, a problem arises from the apparently low probabilities of certain catastrophic accident sequences. In such a case the simple product of the reduction in probability and the consequences of the accident can be seriously misleading. While many people would agree on spending $1 to avoid a mishap causing $10 in damages and occurring with a probability of one-tenth ($1 = $10 × 1/10), their attitude toward a larger accident with smaller probabilities might be different. What amount would people be willing to spend to avoid an accident that would cause $10 billion in damage but would occur with a probability of one-billionth? Would it still be the product of the reduction in probability and the consequences of the accident, $10? More likely, it would be more than $10 for an individual afraid of such a serious accident—or perhaps zero dollars for someone who reasoned that an accident with a probability of one in a billion would never happen anyway. Since many nuclear safety measures change probabilities by a very small amount but might avoid quite severe consequences, this expected-value measure (product of change in probability and dollar consequences) is inappropriate here.

Even beyond the expected-value problem is the difficulty that people might insist on pursuing nuclear safety beyond the point justified by a full reckoning of the damages, pain, and suffering of a nuclear accident. On the dimensions used to measure dread of a technology, nuclear power ranks high. It is new, not old; uncommon, not common; lethal, not merely injuring; involuntary, not voluntary; and potentially catastrophic, as opposed to having noncatastrophic maximum consequences.[33] The public might therefore be willing to pay more for nuclear safety than for the equivalent amount of non-nuclear safety. But how much more? Measuring the benefits of nuclear safety is a difficult task involving many problems of analysis in its own field as well as in others (such as assigning a dollar value to life saving).

Measuring the costs of nuclear safety is somewhat easier. Apart from the usual accounting controversies are two distinct problems: getting information on costs and the possibility that existing methods of achieving nuclear safety are not the most cost effective. Getting information on costs is difficult because the parties that presumably

33. For a discussion and references, see ibid., pp. 10-11.

know the most about costs—the licensees and vendors—have an incentive to overstate costs. If a given safety measure can be portrayed as having excessive costs, then the NRC might not require its adoption. The NRC staff has little reason to overstate or understate costs, but it also has little or none of the detailed operating information necessary for an accurate measure of costs. The question of cost effectiveness of nuclear safety will be examined further in the next section of this chapter.

In comparing costs and benefits of nuclear safety, the analyst is dealing with incommensurables. The costs are dollar sums spent with certainty to promote safety. The benefits are the dollar effects of accidents avoided, but weighted by the low and uncertain probability that the safety system will be called on and will prevent the accident. In important cases, the costs of safety systems will be greater than the simple expected value (probabilities times consequences) of benefits, if only because the probabilities are so low. Therefore the issue of whether greater nuclear safety is justified on cost-benefit grounds will depend crucially on just how high a premium society puts on avoiding nuclear accidents, beyond the expected value of damages avoided. An ounce of prevention is worth a pound of cure, but the proverb offers no guidance when the ounce may prevent a catastrophic accident with an infinitesimally small and vastly uncertain probability.

Since the benefits of nuclear safety measures are so uncertain, it may be worthwhile to pursue reduction of that uncertainty. There is great technical uncertainty, for example, about the merits of some engineered safety features of nuclear power plants. There is little uncertainty about such measures as remote siting of nuclear plants and emergency planning; these measures seem likely to reduce the consequences of an accident even if the worst happens. Making the worst possible outcome more tolerable, the famed "minimax" solution from game theory, may be sensible for some nuclear safety problems.

It is not now possible to say whether the existing level of nuclear safety is too high, too low, or about right on cost-benefit grounds. Such a determination may never be possible. This uncertainty, however, does not rule out the possibility of useful policy analysis. It may be possible to achieve the current level of safety at a lower cost, or— even more promising—to increase the level of safety being achieved without increasing the cost. Such beneficial changes will be possible if safety is not now being achieved cost effectively.

Is Safety Being Achieved Cost Effectively?

In principle, a cost-effective safety program is not hard to devise. The analyst would simply calculate the dollar costs of achieving a given small improvement in safety. Such calculations would be made for all known safety improvements. Then the analyst would choose those safety measures that had lower rather than higher costs for achieving the improvements in safety. A characteristic of the cost-effective safety program would be that, through any avenue of pursuing safety, the standard safety improvement would be achievable at the same cost as for any other avenue of pursuing safety. If some safety methods were more costly at the margin than others, then a reallocation of resources toward effective methods that were less costly at the margin would improve safety without increasing cost.[34]

The NRC has made at least one decision that explicitly violated cost-effectiveness criteria. In setting interim cost-benefit criteria for reducing the low-level radiation released through normal operation of plants, the NRC in effect directed that one effluent be "over-controlled." The standards imply cutting releases of radioactive iodine-131 beyond the point where diverting resources to clean up other radioactive effluents would actually improve radiological health.[35] It has been calculated that the standard for iodine-131 is equivalent to a value of $100 million per statistical life.[36] This value is greatly divergent from the value implied by other NRC decisions, meaning that more health and lives could be saved if resources were allocated away from control of iodine-131.

Although in the matter of iodine-131 effluents the commission's violations of cost-effectiveness principles were explicit, more often the violations are subtle. They result not from a calculation of cost-effectiveness ratios but from the adoption of a working criterion other than cost effectiveness in regulation. The working criterion that appears to have caused the most trouble for the NRC is the goal of visible regulation. Instead of being chosen for their effectiveness, policies have at times been chosen for their visibility.

Visibility and Safety Regulation. The contention is not that NRC staffers are malevolent or incompetent, but that their incentive structure inevitably leads them to value visibility of their safety efforts.

34. See Joseph B. Rivard, "Risk Minimization by Optimum Allocation of Resources Available for Risk Reduction," *Nuclear Safety*, vol. 12, no. 4 (July-August 1971), pp. 305-9. This article is one of a handful in the nuclear literature directly concerned with cost effectiveness.

35. The commission did this by setting cost-effectiveness standards of $1000/man-rem of radiation and $1000/thyroid-rem of radiation, despite the different effects of whole-body and thyroid doses. See the NRC opinion, "Radioactive Material in Light-Water Cooled Nuclear Power Reactor Effluents," Opinion of the Commission, Docket No. RM-50-2, NRC-CLI-75-5 (April 30, 1975).

36. Bernard L. Cohen, "Society's Valuation of Life Saving in Radiation Protection and Other Contexts," *Health Physics*, vol. 38, no. 1 (January 1980), p. 36.

After all, nuclear safety is very difficult to measure. With no day-to-day measurement of the output of nuclear safety, NRC must point to the inputs being devoted to nuclear safety. These inputs—in the form of regulations, inspections, directives, and the like—are quite easily measurable. When called on to demonstrate that it is working to ensure nuclear safety, the NRC has to rely on a show of inputs. Actual nuclear safety is too difficult to show, and the Three Mile Island accident has since 1979 tended to belie assurances of safety based on operating experience.

The difficulty in having direct NRC regulation achieve cost effectiveness is that direct regulation must be centered on the licensee's inputs. NRC can specify such matters as safety systems, shift staffing, and improved technical competence. But it cannot direct that those inputs be combined efficiently without widespread intrusion into the licensee's operations, even assuming that NRC has superior knowledge of how to combine safety inputs efficiently. Nor can the NRC specify a safety output, that the licensee achieve a specific level of safety, when the level of safety is so hard to measure.

The following examples of safety matters where NRC performance has been found lacking can be understood in the context of the need for visibility.

Staff emphasis on safety issues. The ACRS has noted that research tended to track visible safety issues, such as emergency core-cooling, and to pay less attention to mundane concerns that ultimately have a greater effect on safety: "The emphasis on such sophisticated technological questions may have diverted the attention of the ACRS and the NRC Staff from many of the more routine safety-related problems that often precede major accidents."[37]

Man-machine interaction. All the major studies of the Three Mile Island accident indicated that the NRC had given inadequate attention to the human element in nuclear safety.[38] Many factors had interacted to produce the NRC's neglect of operator training, control room design, emergency procedures, and the like. But one factor contributing to the neglect of the human element was that requiring equipment produces a visible measure of promoting safety; the effect of requirements on human behavior is practically invisible. The tendency toward visible regulation as well as the greater en-

37. ACRS, "A Review of NRC Regulatory Processes and Functions," NUREG-0642 (Washington, D.C.: U.S. Nuclear Regulatory Commission, January 1980), p. 4-11.

38. U.S. Nuclear Regulatory Commission, "NRC Action Plan Developed as a Result of the TMI-2 Accident," NUREG-0660 (Washington, D.C.: U.S. Nuclear Regulatory Commission, May 1980), p. I-4.

forceability of requirements for certain equipment and the licensees' own motives for equipment-oriented safety suggest why human factors escaped the attention of the NRC staff.

Licensing versus operating safety. When asked whether existing nuclear reactors are safe, NRC officials often reply by pointing to the long and complex licensing process with its detailed technical reviews and chances for public participation. Licensing, as a visible method of improving safety, has drawn a great deal of attention and resources from the NRC. Day-to-day monitoring of operating experience is less visible and, historically, has drawn less attention. Despite warnings that operating problems were not being given adequate attention, it took the Three Mile Island accident to get resources devoted to this area.[39]

Emergency preparedness. It is clear in retrospect that emergency preparedness was deficient in the pre–Three Mile Island era. Emergency preparedness—dealing with state and local authorities, planning for evacuation—was a subject with little visibility. Had it been more visible, however, it might have revealed inattention rather than attention to nuclear safety. Emergency preparedness was another cost-effective safety measure neglected until after the Three Mile Island accident.

Formalism and licensing. The NRC licensing process has been criticized as excessively formal and legalistic.[40] The casual observer at a licensing hearing is struck by the proportion of time spent in legal maneuvering as opposed to discussion of technical issues. To repeat a question asked of the AEC, does the public interest demand due process or technical expertise?[41] The NRC's interests have demanded an elaborate and visible show of due process, possibly at the expense of a more rational and technically centered licensing procedure.

Resident inspectors. The NRC has made much of its plans to assign resident inspectors to the sites of nuclear facilities.[42] NRC adoption of the resident inspector plan, however, may be as much related to the desire to have a visible NRC presence onsite as much

---

39. The NRC was studying a report on its failure to identify operating problems at the time of the Three Mile Island accident. See NRC, *1980 Annual Report*, p. 90.

40. ACRS, "A Review of NRC Regulatory Processes and Functions," p. 7-8.

41. Elizabeth S. Rolph, *Nuclear Power and the Public Safety: A Study in Regulation* (Lexington, Mass.: Lexington Books, 1979), p. 48.

42. See NRC, *1980 Annual Report*, pp. 139-40.

as to the desire for improved safety. As one inquiry group has noted, no single inspector can assess all the different systems in a large nuclear plant, and the inspector may become a "captive" of the licensee's staff through day-to-day contact.[43]

Other examples of the NRC's pursuit of visible safety, from heavy paperwork requirements on licensees to the NRC's employment of a highly credentialed staff, could be recounted. Still, the pursuit of visible safety does not necessarily contribute to the cost-effective pursuit of actual safety.

Bureaucratic Impediments to Cost Effectiveness. In addition to special problems resulting from the nature of its safety mission, the NRC has some of the problems associated with bureaucratic structures in general, both government and nongovernment. First is the tendency for empire building, both by individuals within the organization and by the organization as a whole. The program offices have been described as independent fiefdoms eager to defend their turf. Moreover, the agency itself comprises many activities that have little reason to be at the NRC. The commission and staff can hardly be blamed for most of these, however, since they were duties passed along by statute. Among other things, the NRC concerns itself with foreign policy (through export licensing that would be better handled by the executive branch), with antitrust implications of nuclear power (as though nuclear power had competitive implications unfathomable by the Department of Justice), and with whether a region served by the utility needs the power to be generated (a responsibility under the National Environmental Policy Act). Evidence suggests that these diverse responsibilities interfere with safety reviews.[44]

A second inevitable tendency of bureaucratic organization is redundancy and waste. In nuclear regulation, redundancy is perhaps best illustrated by the numerous levels at which licensing decisions are made and appealed. The appeals process allows few questions to be decisively settled at any stage of the process. The problem would be worse than it is but for a U.S. Supreme Court decision in 1978 narrowing the grounds on which nuclear licensing decisions may be appealed to the federal courts.[45] The Court ruled that federal judges should not substitute their judgment for that of the NRC, but should overturn agency actions only when the record showed substantial errors in procedure or substance.

43. *Report to the Commissioners*, p. 100.

44. ACRS, "A Review of NRC Regulatory Processes and Functions," p. 7-8.

45. Vermont Yankee Nuclear Power Corp. v. Natural Resources Defense Council, Inc. et al., 435 U.S. 519-558 (1978).

We have seen that there is considerable controversy over just what level of safety is now being achieved at nuclear plants and that there is considerable uncertainty about what level of safety would be appropriate. There is little doubt, however, that nuclear safety is not now cost effective.

What Accounts for the Current State of Nuclear Safety?

The current state of nuclear safety can be attributed to a variety of factors that include the difficulty of the regulatory task itself and past policy mistakes. The problem is not that evil or incompetent people are in important positions, but rather that policies and institutions may have undermined the efforts of even the brightest and most motivated in the NRC and the industry.

Inherent Difficulty of Regulating. First, it should be recognized that regulating nuclear safety is an extraordinarily difficult job under the best of circumstances. In an area with such vast uncertainties, there is often little hard information to go on. The difficulty of nuclear regulation can perhaps be captured in two statements worth elaboration. (1) Conservatism does not ensure safety, and (2) compliance does not ensure safety.

Many nuclear safety decisions require the choice of a key parameter from a wide plausible range. The NRC has tended to choose the parameter value that is least favorable to an attempt to demonstrate safety. The thinking is that if safety can be demonstrated with the "conservative" parameter value—and the true parameter value is probably more favorable—then safety is ensured.

Conservatism does not ensure safety, however, when there is an inadequate basis for even deciding on the range that a parameter could occupy. An example is the NRC's setting of a cost-benefit standard for the amount of damage done by a standard unit of radiation, the "man-rem."[46] If one could only specify a dollar figure for the amount of damage done by a man-rem, it would be easy to decide which measures for cutting the number of man-rems would be justified.

Published estimates showed that the damage done by a man-rem might range from $10 to $980; the commission set a standard slightly more conservative than the published range at $1,000 per man-rem. Surely the commission had erred on the side of safety, if it had erred.

46. See NRC, "Radioactive Material in Light-Water Cooled Nuclear Power Reactor Effluents," p. 283.

Tracing the key figure of $980 per man-rem back to its source is a disturbing exercise. The $980 figure is 1/1,000 of a "value-of-statistical-life" figure of $980,000, based on the grounds that receiving a man-rem of radiation is about 1/1,000 as bad as dying.[47] But how was it decided that a life was worth $980,000? This figure was inferred from the hazardous duty pay of an air force pilot and reflects the air force's judgment during the cold war of how much it should pay pilots for taking on extra risks. The calculation from air force pay was contained in a doctoral dissertation completed in 1963 and was not indexed to account for more than a decade of inflation that elapsed before the NRC ruled.[48] Thus we see that the $1,000 per man-rem standard is based on what the air force was willing to pay pilots during the cold war, a calculation wrenched out of context and not adjusted for the decreased value of the dollar over more than a decade.

The apparent prudence of the NRC in choosing the most conservative value evaporates when the basis of the range of values presented the commission is known. The NRC was actually choosing the most conservative value from a set of wild guesses—a procedure not uncommon in nuclear regulation.[49]

That said, it should be noted that the $1,000/man-rem standard is generally regarded as quite a conservative standard. The point is that, when information for decisions is as incomplete as it is in nuclear regulation, conservatism does not ensure safety.

Compliance also does not ensure safety. The NRC Office of Inspection and Enforcement has promised to make noncompliance more expensive than compliance for licensees, but such a policy still may not ensure safety. Because the licensees are regulated utilities, the question of who really bears the burden of fines is muddied. Depending on practices and accounting methods of state regulators, utilities might be able to pass some of the burden along to rate-payers. Compliance also does not ensure safety if the NRC's regulations are not well designed toward producing safety. If the NRC

47. See Harry J. Otway, "The Quantification of Social Values," in Harry J. Otway, ed., Risk vs. Benefit: Solution or Dream, LA-4860-MS, proceedings of a symposium, November 11-12, 1971 (Los Alamos, N.M.: Los Alamos Scientific Laboratory, 1972).

48. Jack W. Carlson, "Valuation of Life Saving" (Ph.D. diss., Harvard University, 1963). An interlibrary loan request indicated that the NRC library does not have this volume, even though its calculations critically determined the $1,000/man-rem standard.

49. One NRC office coined the term "onageristic" for especially uncertain estimates. A senior program analyst explained that "onager" is a crossword puzzle synonym for "a wild ass" and that some estimates had just that character.

has less detailed or less useful information about a given plant than the utility, the vendor, or the architect-engineer, then NRC regulations might not correspond with cost-effective safety in the plant. The NRC cannot rest on assurances that noncompliance is more expensive than compliance.

Failure to Make Hard Decisions on Risk. The failure to make difficult decisions about risk in society also contributes to the current state of nuclear safety. Recall that Congress has provided virtually no guidance about how much risk, and what nature of risk, can be borne for the benefits of nuclear technology. Since the decision about acceptable risk is a social one, and Congress is the forum in our society for deciding such questions, it is the duty of Congress to provide at least rudimentary guidance. As it stands, there is no social decision-making process on risk. "We make do with disparate efforts of individuals, special interest groups, self-appointed public interest groups, and legislative, judicial, and regulatory systems," is the way that Cyril L. Comar put it.[50]

If Congress has failed to give appropriate guidance, so too has the NRC failed to make tough decisions on risk. It still has no clear policy on how safe is safe enough for a nuclear plant. In the absence of a clearly conceived goal, even a nonquantitative one, the safety goal becomes "just a little safer than it was."

This setting explains the indecision, delay, and redundancy of nuclear regulation.

Diffuse Responsibility for Safety. To a degree unappreciated before the Three Mile Island accident, responsibility for nuclear safety is quite widely diffused. Both within and among the parties involved, there often seems no clear answer to the question of who is responsible. Within the NRC, safety responsibility is diffused over time. Construction permits for plants, for example, have been issued on the grounds that certain safety issues would be resolved later. But when the plant was nearing completion and an operating license was applied for, issues would remain unresolved and there would be pressure to get the reactor into service. Issues might be deferred from the construction permit phase to the operating license phase, but never resolved.[51] A diffusion of responsibility for safety among

50. Cyril L. Comar, quoted in ACRS, "An Approach to Quantitative Safety Goals for Nuclear Power Plants," p. 33.

51. Stanley M. Gorinson, chief counsel, "Report of the Office of Chief Counsel on the Nuclear Regulatory Commission," submitted to the President's Commission on the Accident at Three Mile Island (October 1979), p. 41.

the staff of the reactor vendor for Three Mile Island prevented the transmission of information that might have forestalled the accident.[52] Among the staff of the operating utility at Three Mile Island, nobody had full responsibility for safety.[53]

Responsibility for safety is again diffused among the various institutions involved in nuclear power. The NRC has a statutory mandate to protect public health and safety, but the NRC holds the owner of the plant ultimately responsible.[54] Meanwhile, the plant owner has less expertise than the vendor who supplied the plant or the unlicensed architect-engineer who built the plant.[55] Exemplifying this diffusion of responsibility, the owner of the Three Mile Island plant filed a $4 billion claim under the Federal Tort Claims Act against the NRC. The contention was that faulty NRC regulation allowed the Three Mile Island accident to happen.[56]

The nuclear industry is certainly not alone in having distinct entities dividing safety responsibilities. Why, then, has the nuclear industry been plagued so with difficulty in coordinating safety responsibilities?

Economic Disincentives to Cost-Effective Safety. A prominent reason for the nuclear industry's special difficulties in coordinating safety responsibilities is that it lacks an important economic stimulus to safety present in other hazardous industries—liability for damages in the event of an accident. Accident liability makes it somebody's responsibility to be concerned about safety, instead of being nobody's responsibility. The nuclear industry, as noted in chapter 1, is unique in the lack of accident liability. What is more, the small liability that exists has been placed on the party with the least expertise about nuclear safety—the operating utility. The reactor vendor with all its knowledge of safety cannot be sued for offsite damages in the event of a serious nuclear accident. The vendor, therefore, has much less incentive to use its knowledge to promote safety than it would have under the customary stricter liability arrangements.

Studies of the nuclear industry accept as immutable that improved safety is not in the industry's financial interests. One reads

---

52. Stanley M. Gorinson, chief counsel, "Report of the Office of Chief Counsel on the Role of the Managing Utility and Its Suppliers," submitted to the President's Commission on the Accident at Three Mile Island (October 1979), p. 137. Cited hereafter as "The Managing Utility."

53. Ibid., pp. 66-74.

54. ACRS, "A Review of NRC Regulatory Processes and Functions," p. 5-1.

55. Report to the Commissioners, p. 100.

56. See NRC, 1981 Annual Report, pp. 150-51.

of "the NRC's reliance on the industry to regulate itself despite the industry's financial disincentives to do so" and of industry willingness "to take risks in excess of those the public was willing to accept."[57]

What these studies have not recognized is that safety is not in the financial interest of the industry precisely because it faces little liability in the event of an accident. Therefore, the industry's financial disincentives to safety should not be lamented as an unfortunate fact of the world, but should be regarded as something that can be changed in the interest of safety.

There are numerous ways in which the industry's severely limited liability affects safety decisions and the relationship with the NRC. When the licensee's preferred safety level is well below that of the NRC, it means that the licensee has to be dragged unwillingly into accepting safety improvements. One inquiry into nuclear safety referred to experience showing that the industry will be slow to move unless the NRC lays out minimum requirements and a deadline.[58] Further, when limited liability places the licensee's preferred safety level below the NRC's, the licensee will comply with no more than the minimum NRC regulations. Exhortations to go beyond NRC requirements will be useless. The matter of safety incentives also affects reporting. Vendors and licensees have reasons not to identify and report safety problems, since the NRC may require costly backfitting.[59] The backfitting would have high costs but small benefits to the firm since it would bear little or no liability in the event of a serious accident. Limited liability has made the nuclear regulatory process much more contentious than it needs to be, since it makes the licensee's preferred safety level diverge so from the NRC's.

Perhaps more important than what limited liability does to safety today is what limited liability does to safety innovation for the future. Were liability greater, safety research lowering the expected value of accident claims would have a high payoff. With liability so limited, safety research has a smaller payoff for vendors and ultimately for licensees. This is not to say that limited liability is the only source of disincentive for safety innovation. On the contrary, several features of NRC regulation discourage innovation. The ACRS has concluded that even when safety innovations can accomplish substantial cost savings, the problems brought on by delays for review of the new design features usually discourage innovation.[60] Also, an NRC chair-

---

57. "The NRC," p. 3; Rolph, *Nuclear Power and the Public Safety*, p. 163.
58. *Report to the Commissioners*, p. 128.
59. Ibid., p. 164.
60. ACRS, "A Review of NRC Regulatory Processes and Functions," p. 5-4.

man has testified that vendors are reluctant to suggest safety improvements for fear that the NRC will require that they be supplied to all like plants.[61]

Still another potential disincentive for innovation is the use of "as low as reasonably achievable" requirements. These requirements take the form of cost-effectiveness ratios; if a given radiation effect can be avoided by spending less than a specified amount of dollars on safety, then the safety measure must be taken. If the cost of avoiding the radiation effect is more than the specified amount, then the safety measure need not be implemented. Under these circumstances, vendors are assured that if their research efforts find a cost-effective way of reducing radiation, they will have to provide it on all future units and perhaps backfit existing units as well. The nuclear industry has substantial incentives not to do safety research of the amount and quality that it might, and the NRC's safety research programs can only be a partial substitute for industry research.

Federal regulation, in addition to that of the NRC, and state public utility regulation may also be disincentives to safety. Federal income tax regulations and regulation of wholesale power rates might cause utilities to rush through preoperational testing to get the plant on line quickly.[62]

The responsibility for nuclear safety is mostly a federal one, especially in light of a 1972 case in which the Supreme Court upheld a ruling that the federal government has almost complete authority on operation and construction of nuclear plants.[63] The states, though, have the right to regulate electric rates. This rate-making authority can have important consequences for safety and for the cost effectiveness with which it is achieved.

Economists have long recognized that rate-of-return regulation, as practiced by the states, tends to cause utilities to use inefficiently large amounts of physical capital in production.[64] Under rate-of-return regulation, the utility can keep more of its profits if it has a large amount of physical capital—such as generators and transmission lines—in service. It is easy to deduce a parallel bias in the production

61. "The NRC," p. 101.

62. *Report to the Commissioners*, p. 163. The staff of the President's Commission on the Accident at Three Mile Island was unable to determine whether tax considerations had been a factor in that incident. See "The Managing Utility," p. 53.

63. Northern States Power v. AEC, 447 F. 2d 1143 (8th Cir., 1971), upheld by the Supreme Court at 405 U.S. 1035 (1972).

64. See Harvey Averch and Leland L. Johnson, "Behavior of the Firm under Regulatory Constraint," *American Economic Review*, vol. 52, no. 5 (December 1962), pp. 1052-69.

of safety, a bias toward installed equipment and away from non-capital-intensive methods of producing safety, such as operator training. The traditional neglect of human factors in nuclear safety is consistent with both the financial interests of the licensee and the NRC tendency toward visible regulation.

State regulation can also affect safety in its choice of which costs can be passed along to ratepayers and which cannot. If, for example, safety reasons dictate shutdown or reduced power at a nuclear plant and the utility cannot pass along the cost of replacement power, the utility will think twice before taking the safety measure. Further, safety features not required by the NRC may not be viewed as "used and useful" by state regulatory commissions and might not be calculated into the rate base.[65] These examples suggest that there may be an important interaction between nuclear safety regulation and state public utility regulation, which has long been ignored by the NRC and its staff.

Throughout this examination of the results of nuclear regulation, one theme has emerged: the separation of knowledge about nuclear safety from the incentive to apply that knowledge. As matters stand, there is no locus of decision making in the nuclear industry where knowledge and incentive are combined. Great amounts of resources may be going into nuclear safety, but they are not being used cost effectively. Fundamental improvements will await a better definition of what we want nuclear regulation to accomplish and a better union of knowledge about safety with the incentive to be safe.

65. *Report to the Commissioners*, p. 163.

4

Initiatives for Improving
Nuclear Regulation

The preceding chapters have shown that the uncertainties involved in nuclear safety and nuclear regulation are vast. Just how much safety is now being achieved is uncertain, as is the justifiability of the safety level on benefit-cost grounds. Congress and the NRC have failed to make some difficult decisions on risk, and the knowledge of how to achieve nuclear safety has remained diffused among a number of parties. As a result, nuclear regulation has drifted along in redundancy and indecision, causing more and more resources to be devoted to safety, though probably not cost effectively.

Nuclear regulation might have continued in just this fashion for years had it not been for the Three Mile Island accident. The accident and the apparent ineptitude with which NRC and licensee alike responded set off warning signals. Outside inquiries into the accident found serious deficiencies in specific nuclear safety practices and in regulation generally. Although reforms were quickly implemented in the specific safety areas identified, larger changes in the conduct of regulation have not been made.

Whatever shock and exhortation can do to improve safety regulation has now been done. The Three Mile Island accident will not happen again, and there may even be an overkill in responding to the particular safety problems the accident revealed.[1] The obvious steps toward improving safety that can be taken without new legislation— such as requirements for more remote siting and emergency planning —have also been taken or are in progress.

New safety challenges will arise in the future. Will they be better handled than Three Mile Island? The answer lies in whether the institutions and practice of nuclear regulation will have been reformed and improved at the time of the next serious challenge.

1. See the testimony of Stephen Howell in U.S. Congress, House Committee on Interior and Insular Affairs, Subcommittee on Energy and the Environment, *Hearings on H.R. 6390* et al. (96th Cong., 2d sess., March 13, 14, 18; May 12, 1980), p. 46. Hereafter cited as 1980 *Hearings*.

After all, even the beneficial effects of the Three Mile Island accident in promoting safety awareness and in emphasizing that accidents can happen in the absence of vigilance will fade with time. Only if institutions and practices have been improved can there be confidence that safety challenges will be met successfully.

Changes should proceed on three main fronts if confidence in nuclear safety is to be restored. First, the NRC's own house should be put in order. Second, social decisions should be separated from technical ones in nuclear regulation. Third, the now-deficient incentives for safety should be corrected and the NRC's style of regulation correspondingly changed.

Put the NRC's House in Order

The NRC has done what it can to respond to the Three Mile Island accident. Unfortunately, the response has not included significant streamlining of its management structure. Although the roles of the chairman of the commission and the executive director for operations have been strengthened, there are still two competing (and occasionally conflicting) centers of influence. The new units formed after the Three Mile Island accident, such as the operational data office and the human factors safety division, were simply tacked onto the existing structure. The NRC organizational chart has increased in complexity, but there is little indication that the agency is being managed more coherently. The shuttling of former Chairman Joseph Hendrie in and out of the chairmanship during 1980–1981 certainly did not help the continuity of leadership.

If the NRC's house is to be set aright, it must be done within the NRC. Detailed guidance from Congress about just how to achieve nuclear safety is not the answer, and some bills specifying particular safety measures have been lost in the post–Three Mile Island crush.[2] This is for the better. The answer is organization of the NRC so that the public can have confidence in its conduct of regulation, not a congressional mandate of specific safety procedures and staffing requirements.

Much of the disorganization of the NRC staff, since the founding of the agency in 1975, can be traced to the de facto existence of "two NRCs": the one in downtown Washington associated with the commission itself and the one in Maryland associated with the executive director for operations. The reorganizations to date have not touched the dual nature of the staff, and therefore have not relieved the unmanageability of operations. Further, when consolidation (man-

2. For examples of such measures, see 1980 *Hearings*, pp. 301-433.

agerial and physical) of the two staffs is considered, it becomes obvious that only the strongest of agency chief executives could command the loyalty of the consolidated staff.

As long as there are five separate commissioners with their own constituencies in the staff, the NRC will continue to be an unmanageable organization. Both the President's Commission on the Accident at Three Mile Island and the NRC's own Special Inquiry Group recommended that the five commissioners be replaced with a single administrator.[3] It is easy to dismiss such recommendations as merely symbolic. The parallel existence of two competing staffs, however, is not a symbolic threat but a real one to the successful management of the agency. There is but a small chance that any one person—no matter how talented—can step in as NRC chairman and manage the staff as it is now organized. The NRC must put its house in order from within, but this will be virtually impossible without a strong chief executive.

Separate Technical Decisions from Social Decisions

Congress has refused to make some hard decisions on what kinds of risks we as a society are willing to accept in return for the benefits of nuclear technology. Legislative guidance to the NRC, and to the AEC before it, has consisted of a suggestion that the agencies protect public health and safety, as though nuclear technology could only be "safe" or "unsafe." There has been no guidance on how far to go in the pursuit of nuclear safety, inherently a matter of degree.

In the absence of guidance from Congress about the inherently social question of acceptable risk, the NRC—constituted as a technical decision-making body—is preparing to make a key social decision. The commission proposes to specify "how safe is safe enough."[4] This effort, if successful, would provide a safety standard that could be used in licensing and other decisions. Then decisions on specific cases could be narrowed to the issue of whether the license or rule proposed would be consistent with the safety goal. As one NRC commissioner expressed the potential for a safety goal:

> Such a statement could help bridge the gap between the law's broad but vague instructions—"adequate protection to

3. John C. Kemeny, chairman, *The Need for Change: The Legacy of TMI*, Report of the President's Commission on the Accident of Three Mile Island (October 1979), p. 61; Mitchell Rogovin, director, Special Inquiry Group, *Three Mile Island: A Report to the Commissioners and to the Public*, NUREG/CR-1250) (Washington, D.C.: U.S. Nuclear Regulatory Commission, January 1980), pp. 115-17.

4. NRC, "Proposed Policy Statement on Safety Goals for Nuclear Power Plants," *Federal Register*, vol. 47, no. 32 (February 17, 1982), pp. 7023-28.

the health and safety of the public"—and the specific day-to-day decisions faced by the Commission and its staff, so that every new decision does not require a return to the Act, the fundamentals of mathematics, and the laws of nature for a redetermination of how much protection is "adequate."[5]

The commission considered six proposals of quantitative safety goals, each as a candidate for "safe enough."[6] Its proposed safety goal contains two qualitative statements and three kinds of numerical goals:

Qualitative statements: (1) Individual members of the public should be provided a level of protection from the consequences of nuclear power plant accidents such that no individual bears a significant additional risk to life and health.

(2) Societal risks to life and health from nuclear power plant accidents should be as low as reasonably achievable and should be comparable to or less than the risks of generating electricity by viable competing technologies.

Numerical goals: (1) Individual and societal mortality risks: (a) The risk to an individual or to the population in the vicinity of a nuclear power plant site of prompt fatalities that might result from reactor accidents should not exceed one-tenth of one percent (0.1%) of the sum of prompt fatality risks resulting from other accidents to which members of the U.S. population are generally exposed. (b) The risk to an individual or to the population in the area near a nuclear power plant site of cancer fatalities that might result from reactor accidents should not exceed one-tenth of one percent (0.1%) of the sum of cancer fatality risks resulting from all other causes.

(2) Benefit-cost guideline: The benefit of an incremental reduction of risk below the numerical guidelines for societal mortality risks should be compared with the associated costs on the basis of $1,000 per man-rem averted.

(3) Plant performance guideline: The likelihood of a nuclear reactor accident that results in a large-scale core melt should normally be less than one in 10,000 per year of reactor operation.[7]

5. Separate Views of Commissioner Gilinsky in NRC, "Proposed Policy Statement on Safety Goals for Nuclear Power Plants," p. 7027.

6. See NRC, "Toward a Safety Goal: Discussion of Preliminary Policy Considerations," NUREG-0764 (Washington, D.C.: U.S. Nuclear Regulatory Commission, Office of Policy Evaluation, March 1981), pp. 42-45.

7. NRC, "Proposed Policy Statement on Safety Goals for Nuclear Power Plants," pp. 7024-26.

While one certainly can quibble with details of the proposed safety goal, the principle of establishing some safety goal may well be more important than the specific goal adopted. In the absence of an articulated safety goal, the NRC might continue to drift along toward a goal of "more safety" with little concrete idea of its objective. To be sure, the establishment and refinement of an appropriate safety goal is a difficult task. It may be that the proposed set of goals is too abstract, as NRC Commissioner Victor Gilinsky believes.[8] But the job of formulating a meaningful safety goal has begun. One can only wonder why a clear articulation of safety goals was not begun much earlier in the history of the NRC or its predecessor, the AEC.

The dichotomy between social and technical decisions suggests a corresponding dichotomy in the treatment of public participation in nuclear regulation. On social decisions, public participation is vital. The willingness of a society to accept risk, and the nature of the risks society wants to accept, are inherently social matters. They are to be decided by each citizen and through political processes responsive to citizens. Further, technical expertise gives one no particular license to make social decisions. Because public participation is vital in the making of such decisions, provisions for public funding of intervenors are entirely appropriate.

Once the social decision is made, however, and the nature of acceptable risk determined, the nuclear issue becomes a technical one. Does the rule or facility conform with the socially determined acceptable risk? The kinds of skills for answering such questions are inherently technical. Technical questions call for technical answers, and here the ordinary citizen's views are much less valuable. Public participation at this point should be narrowly circumscribed to the technical issues at hand, and only technically qualified opinions need be admitted.

In separating technical from social issues in licensing, the ACRS has made a useful suggestion. The proposal is to constitute an authority during the licensing process to make a legally binding determination of the spectrum of risks imposed by a given plant.[9] Then the risks imposed by the plant would not be a subject for redundant debate at multiple levels of appeal.

If the scope of licensing decisions is to be made more narrow and technical, certain nontechnical parts of the NRC empire need to be farmed out to more appropriate agencies. Obvious candidates for

8. Separate Views of Commissioner Gilinsky in NRC, "Proposed Policy Statement on Safety Goals for Nuclear Power Plants," p. 7027.

9. ACRS, "An Approach to Quantitative Safety Goals for Nuclear Power Plants," p. 75.

removal would be the NRC's export licensing authority, antitrust review, and plant-by-plant review of the needs of the region for power.

When agency decisions become more technical, then it may be possible to "decide to decide" certain matters just once. The existing two-stage license review, for example, could be replaced with a one-stage process in which the utility would submit information detailed enough for an operating license before construction began. The NRC would decide to decide in advance whether the plant could be constructed and operated within safety guidelines. Having decided, the NRC would not reopen this issue, though of course it would continue surveillance of construction and operation to see that preconstruction plans were carried out. In the absence of errors in the initial analysis or substantial new evidence, the utility could count on not having to backfit safety equipment not required at the time of licensing.

Proposals to replace the five-member commission with a single administrator complement the separation of technical and social issues. A collegial body may be appropriate for making social decisions because of the diversity its members contribute, but in the efficient management of technical decisions, a collegial body is less useful than a single administrator with the appropriate authority.

Correct the Incentives for Safety

It is useful to think of approaches to nuclear regulation as lying along a continuum. At one end is regulation by direct mandate of design and operating procedure, extending perhaps to government ownership and operation of nuclear facilities. At the other end is regulation through an incentive system, with built-in rewards for being safe and built-in penalties for being unsafe, and licensees left on their own to produce safety.

The NRC's regulatory approach has been heavily tilted toward direct mandate of design and operating procedure. What is more, there have been proposals in Congress to involve the NRC even more intimately in plant operation, to the point of having NRC personnel start up and shut down reactors. So far, the NRC has avoided this approach because it would relieve the operating utilities of their responsibility for safety.[10] Indeed, under regulation by direct mandate, any move by the NRC to assume more responsibility for safety might simply displace licensees' existing efforts to ensure safety. The answer is not to have the government take over the plant, but to get the licensee to run it competently.

10. ACRS, "A Review of NRC Regulatory Processes and Functions," NUREG-0642 (Washington, D.C.: U.S. Nuclear Regulatory Commission, January 1980), pp. 7-4–7-5.

A promising direction for the NRC to move, then, is along the continuum toward a greater reliance on incentives. The NRC has not yet seriously studied the use of incentive systems for enforcing nuclear safety, though it published in 1978 a massive report on the subject, "A Study of the Utility of Incentive Systems for Nuclear Licensees."[11] The study, however, is seriously incomplete and contains fundamental misunderstandings. In discussing existing incentives for safety, the study asserts that "outsiders can sue for damages" and does not recognize the Price-Anderson Act's severe limitation of that right to sue. Indeed, the study does not mention the Price-Anderson Act when it considers disincentives for safety.[12] The study, in total disregard of evidence on electricity demand and state regulatory practice, states that "compliance costs can be recouped and the higher product prices do not threaten sales."[13] The question of who really bears compliance costs, a complex issue worthy of study, is assumed here not to be a question at all. The "incentive systems" studied were not true incentive systems, in most cases, but were instead direct regulation strategies such as changing the frequency of inspections and conducting diagnostic inspections more frequently.

The study evaluated enforcement alternatives through responses to questionnaires sent to NRC and licensee personnel. It chose "promising" alternatives by finding those that were supported by the NRC or licensees or both, and were not opposed by either.[14] Consequently, the alleged incentive systems emerging from this study were a set of weak measures ranging from sponsoring regional seminars on safety to having inspectors make positive statements when they found a well-run operation. Not surprisingly, increased civil penalties for safety violations were not among the incentives recommended. Increased liability for accidents—an obvious incentive for safety—was not among the alternatives evaluated. The study did usefully note, however, that the NRC's resources are too limited to provide for comprehensive direct inspection.[15]

It is time to design and implement an incentive-based approach to nuclear regulation, keeping in mind the rationale for regulation. Although there is no rigorous legislative rationale for nuclear regula-

11. U.S. Nuclear Regulatory Commission, "A Study of the Utility of Incentive Systems for Nuclear Licensees," NUREG/CR-0387 (Washington, D.C.: TRW Energy Systems Planning Division and U.S. Nuclear Regulatory Commission, August 1978).

12. Ibid., pp. 46-47.

13. Ibid., p. 47.

14. Ibid., p. 103.

15. Ibid., p. 39.

tion, as discussed in chapter 1, there is an economic one: the costs of an accident to society may be greater than the costs to the licensee, leading the licensee to underinvest in safety. In other words, the licensee's incentives for safety are deficient. The solution is to correct those incentives and to issue supplemental regulations in those areas where incentives will not work. Where incentives do work, the licensee should be left to its own methods of producing safety. Chauncey Starr claims, for example, to have identified instances in which the licensee would—out of self-interest—seek a safety standard orders of magnitude higher than necessary to protect the public.[16]

A two-pronged strategy for improving nuclear regulation emerges from this discussion: (1) correct the incentives for safety, and (2) construct a regulatory program that complements the newly corrected incentives. In those areas where the licensee's self-interest leads to adequate safety, the regulatory role can be quite limited. Safety programs that protect the public but not the licensee, such as remote siting of reactors, then become the proper focus of a leaner, more concentrated regulatory program.

How can the incentives for safety be corrected? The most obvious answer is to remove the current disincentive to safety provided by the existing form of the Price-Anderson Act. If a nuclear utility faced the possibility of large claims from accident victims—just as the airline and chemical industries do—then it would face rewards for safety and penalties for unsafe operation. Normal liability arrangements would provide the nuclear industry with an incentive to produce safety in fact rather than the safety in appearance all too often required by NRC regulations. Normal liability arrangements would also make the job of the NRC easier, by narrowing the gap between the licensee's preferred level of safety and the efficient level.

Limitations of These Initiatives

There are, however, limitations to these initiatives to reform nuclear regulation. If increased liability is considered in isolation with all other nuclear institutions assumed to remain the same, it can reasonably be seen as having little effect. Increased liability must not, therefore, be viewed as a cure-all in itself. The purpose of increasing liability is to promote safety-enhancing change in nuclear institutions

16. Chauncey Starr, "Risk Criteria for Nuclear Power Plants: A Pragmatic Proposal," unpublished, American Nuclear Society/European Nuclear Society International Conference, November 16-21, 1980. Available from U.S. Nuclear Regulatory Commission Public Document Room, 1717 H Street, N.W., Washington, D.C. 20555.

—change such as greater technical expertise among utilities or larger roles for reactor vendors. This is not a matter of greater liability versus alternatives such as greater technical expertise, but greater liability as a stimulant to acquiring technical expertise. Greater liability provides the economic environment in which greater technical expertise becomes indispensible.

It remains possible that liability could be increased, however, with no response from nuclear institutions. But uncertainty about how nuclear institutions respond to liability is surely an argument in favor of ending the radically different treatment of the nuclear industry. No other industry is so completely shielded from damage claims in the event of an accident. Such a policy could only be proper if there was near-certainty that nuclear firms do not respond to lower liability with lesser safety. It is hardly proper in the presence of the vast uncertainties of today.

A final qualification to the proposals in this chapter is that they are somewhat narrow in scope. They concern what the NRC and Congress could do to improve the regulation of nuclear power. They might have little effect, for example, on licensing time for new facilities. As Linda Cohen has shown, long licensing time "is not an artifact of bureaucracy and red tape."[17] Further, nuclear power has many problems other than regulation—high interest rates and slowed growth of electricity demand, to name two. The initiatives proposed above would not guarantee a substantial increase in nuclear power generation. Nor should they. What they can do is to guarantee that utilities considering nuclear generation will face a consistent and rational set of regulations, instead of the current tangle. If utilities facing a consistent and rational set of regulations decide not to order nuclear capacity, there should be no cries of regulatory failure.

Prospects for Implementation

The prospects of the NRC putting its house in order are not good, given the existing organizational structure. As noted above, it would take an extraordinarily charismatic NRC chairman to unify operation of the two separate staffs and layers of organization. Even if such an individual could be found and agreed to serve, improved organization might last no longer than that person's tenure. If the NRC is ever to have coherent management on a continuing basis, it will have to be under a single administrator as recommended in the reports of the

17. Linda Cohen, "Innovation and Atomic Energy: Nuclear Power Regulation, 1960-Present," *Law and Contemporary Problems*, vol. 43, no. 1 (Winter-Spring 1979), pp. 67-97, and esp. p. 96.

president's commission and the NRC Special Inquiry Group. To replace the commission with a single administrator would require legislation, and such a move is said to have "few friends on the Hill."[18] The impetus for such a change also may have been blunted by a reorganization in 1980 that clarified the roles of the NRC chairman and executive director for operations.

More optimism is justified in the NRC's efforts to separate social and technical issues in licensing and regulation. The NRC's proceedings on "how safe is safe enough" have made significant progress. If the NRC's final safety goal is endorsed by Congress through the oversight function, then the NRC may find itself not having to redecide the acceptability of nuclear power in every proceeding. Having neglected its responsibility to provide broad guidance on the acceptability of risk, Congress surely can provide a simple endorsement or rejection of the safety goal adopted by the NRC.

On correcting safety incentives, prospects are mixed. First, even in the absence of any statutory limitation on liability, there would be a de facto limit on the licensee's accident liability. The licensee could lose only as much as its corporate worth before declaring bankruptcy; corporate worth of even the largest utilities is below the maximum damages of some potential accidents. Regardless of what Congress does, then, the utility will stand to lose less than the public for certain catastrophic accidents.

Congress could repeal the Price-Anderson limit on liability, so that liability would go from $560 million (the current figure) to the corporate worth of the utility. Such treatment would parallel that of other industries using dangerous technologies, who know that their corporate worth is potentially subject to liability claims. The Price-Anderson liability limit confers a sizable subsidy, however, and old subsidies are hard to eliminate. Outright repeal of the Price-Anderson liability limit would certainly face tough going in Congress.[19]

Meanwhile, the $560 million liability figure has become increasingly untenable. This figure was set in 1956–1957 and was never adjusted upward, despite the fact that inflation alone cut the value of the coverage to less than half its original value. Further, the reactors being designed in the 1950s were smaller, produced less radioactive material, and were located farther from population centers than more recent reactors. The real liability level as a proportion of possible damages has shrunk markedly.

Perhaps recognizing the vulnerability of the $560 million figure, the industry has advanced a proposal that would result in coverage

18. 1980 *Hearings*, p. 36.
19. Ibid., p. 15.

of $1 billion for each accident, but would retain features of the Price-Anderson Act that immunize vendors from lawsuits and protect utilities from claims beyond the limit.[20] Since the industry has advanced this compromise, it seems likely that some legislation increasing the amount of liability can be adopted. The plan raising liability to around $1 billion, however, would not restore the level of liability even to its 1956–1957 level. Further, the absolute limit on liability for utilities and the exemption from offsite liability for vendors and architect-engineers would continue to promote a diffusion of responsibility for safety and knowledge of safety.

To summarize, a complete reform package for nuclear regulation would include streamlining the NRC into a single-administrator agency with an emphasis on deciding technical questions, not social ones. The rest of the package would include removal of the current legal disincentives to safety, with the newly reorganized NRC conducting a well-focused inspection and enforcement program designed to protect the public. In those areas where the licensee would be hurting only itself by poor operation, the NRC would not intervene. This package should not be viewed as punitive to the nuclear industry. In fact, it could be seen as a reasonable bargain: the industry gives up its extraordinary immunity from damage claims and gets unobtrusive, streamlined, incentive-based regulation.

Separation of technical and social issues in nuclear regulation is likely to be helpful, independent of whatever other changes are undertaken. But streamlined regulation and increased liability must be seen as strict complements. Simply adding increased liability onto the existing hodgepodge of regulation would not be helpful. Nor would it be prudent to cut back on safety regulation in the absence of proper incentives for safety.

H. R. 6390 and Its Successors

A discussion of specific legislative proposals should make the recommended changes in nuclear regulation seem more concrete. Such proposals are numerous, but the most comprehensive one—and one of the few addressing safety incentives—is H.R. 6390, introduced January 31, 1980, by Representative Morris K. Udall of Arizona. Hearings were held on the bill in March and May of 1980, but it went no further. Parts of it not dealing with regulatory reform were reintroduced as H.R. 8378 and ultimately were included in the Low-Level Radioactive Waste Policy Act, Public Law 96-573. Thus the original H.R. 6390 remains the most comprehensive model legislation to date.

20. Ibid., pp. 52-53.

The bill's first title, relating to plant siting and safety, declares a three-year period of extra caution in nuclear licensing. It requires the NRC to certify that any new plant licensed is consistent with the safety findings of official inquiries into the Three Mile Island accident. The second title provides a timetable for resolving matters related to nuclear waste disposal, on which the Department of Energy has lead responsibility and NRC has a role in ensuring safety. Title III addresses, among other things, public funding for intervenors, the "need for power" determination, and modification of the Price-Anderson Act. Title IV sets up a Nuclear Safety Board, which in effect would oversee the NRC.

Much of the bill's first title has already been put into effect by NRC regulation; since Title II deals with licensing an Energy Department effort, it is not of direct interest in the reform of nuclear regulation. Titles III and IV contain measures broadly consistent with this study's recommendations and, in any event, would change the nature of nuclear regulation considerably.

The Udall bill proposes to begin to streamline regulation and to separate technical from social decisions. The bill delegates to state governments, for example, the determination of whether the power from a proposed plant is needed. This is surely an issue the NRC has little reason to be deciding. The Udall bill also promotes a separation of technical and social issues in providing for early approval of nuclear sites as consistent with safe operation of a nuclear facility. With the state having certified the need for power and the site approved, the licensing decision for a specific facility could be decided on technical grounds.

In recommending a separate Nuclear Safety Board, the Udall bill is setting up regulators to regulate the regulators. A preferable course would be to reform the NRC so that confidence could be placed in its safety determinations. Any future accident on the order of Three Mile Island would presumably be investigated by authorities separate from the NRC, such as the presidential commission that investigated the accident at Three Mile Island.

The Udall bill provides for funding of public participation in NRC proceedings, including both general decisions arrived at in rule making and specific decisions on particular plants. The bill, therefore, does not reflect an appreciation of the differing values of public participation in social and technical decisions. Funding intervention in the licensing of specific facilities could well mean promoting delay. Funding intervention in general rule makings, however, could bring the NRC valuable points of view it might not otherwise hear.

Perhaps the most important provision of the Udall bill would increase the level of coverage for nuclear accident victims to around $5 billion. Liability remains strictly limited to the utility's coverage and does not extend to vendors and architect-engineers.

The source of the increased coverage is twofold. First, the bill increases the amount of coverage now provided by a joint arrangement among nuclear utilities. Under the current terms of the joint arrangement as enacted in the 1975 amendments to the Price-Anderson Act, all utilities with nuclear units help cover the damages caused by any one unit. Following an accident, utilities could be assessed up to $5 million per nuclear unit to cover damages. With seventy-five reactors operating, this arrangement provides a fund of $350 million available to be assessed after a serious accident. The Udall bill would increase the amount that a nuclear utility could be assessed after an accident to $20 million per unit, increasing the total that could be assessed to $1.5 billion.

The second source of increased coverage under the Udall bill is increased insurance written by the government. Since the adoption of the Price-Anderson Act, the government has written just enough coverage to keep total funds of $560 million available in the event of a nuclear accident. In 1957 when other insurance amounted to $60 million, the government provided $500 million. As insurance other than that provided by government coverage has increased, the government has reduced the level of its insurance to less than $60 million. The Udall bill would expand government coverage to $3.5 billion, so that total funds available to accident victims would be around $5 billion.

Although the increase in liability from $560 million to $5 billion may appear to be large, safety incentives would not be significantly affected. The greater part of the increase would be in the form of government insurance, which does not vary from plant to plant. Plants at remote sites and with good safety records would pay the same amount per megawatt as would plants in densely populated areas with bad safety records. Thus the increase in liability covered with flat-rate government insurance would not provide the safety incentive that would be provided by self-insurance or private insurance.

The $1.5 billion provided by increased after-the-fact assessments ("retrospective premiums") would help restore safety incentives. Since all utilities would have a stake in improving the safety of the less safe ones, joint arrangements to increase safety might become more effective. The Udall bill, in summary, would increase safety incentives, but not by as much as the increase in liability from $560 million to $5 billion might imply.

It is easy to argue with specific points in the Udall bill. The premise of the bill, however, is fundamentally sound: some decision on nuclear power is better than total indecision. If the United States is to forgo nuclear power, it is high time to be developing alternative energy sources. If the United States is to have nuclear power, it is high time to reform regulation and liability so that the public can have confidence in the practice of nuclear regulation.

Recent Developments

Since fall 1981, three significant developments in nuclear regulation have occurred. Unfortunately, each reflected familiar approaches to regulation that have served so poorly in the past.

First, the NRC took another step away from streamlining and simplifying its organizational structure. A new position, deputy executive director for regional operations and generic requirements, was created and tacked on to the dual-staff structure, which remains intact.[21] NRC insiders say the holder of such a position, a virtual czar, will be yet another independent source of power in an agency that already suffers from too many independent power centers.

Second, the NRC has implemented a regulatory streamlining proposal that shows it continues to operate oblivious to the possibilities of more incentive-based regulation. The NRC has eliminated the existing review of a utility's financial qualifications to operate a nuclear power plant.[22] The NRC's published rationale does not recognize that this move could foreclose effective increases in corporate liability, by allowing the formation of licensees with little net worth other than nuclear facilities. Thus the NRC is failing to recognize how today's regulatory strategy could eliminate one of tomorrow's options. This is especially ironic since the NRC is committed to study safety incentives with an eye toward future applications.[23]

Third, in its release of Sandia study estimates of accident consequences (discussed in chapter 3), the NRC took actions which invited its portrayal as covering up unfavorable results. Even after worst-case calculations had been released by a House subcommittee and widely reported in the press, the NRC issued a report which did not include worst-case results but instead concentrated on mean or average results. With senior NRC officials releasing and defending

21. See NRC Press Release 81-170, October 16, 1981, on file at the NRC Public Document Room.

22. NRC, "Elimination of Review of Financial Qualifications of Electric Utilities in Licensing Hearings for Nuclear Power Plants," *Federal Register*, vol. 42, no. 62 (March 31, 1982), pp. 13750-13755.

23. See "NRC Action Plan Developed as a Result of the TMI-2 Accident," NUREG-0660 (Washington, D.C.: U.S. Nuclear Regulatory Commission, May 1980), p. V-5.

such a report, it was inevitable that the public image would be that of an agency trying in vain to play down the danger. Matters of image are important when an agency needs to build its public credibility as the NRC does.

Although none of these events is major in itself, taken together they indicate business as usual at an agency where the usual business has been unsatisfactory.

5
Summary and Conclusions

Nuclear regulation is sometimes cited as an example of regulatory failure, which implies personal failures by regulators at the NRC and at the AEC before it. This is too facile an explanation. If there is a regulatory failure in nuclear power, it is a product of diverse influences ranging from the inherent difficulty of the NRC's mission to the unwillingness of Congress to make some hard decisions about risk.

Regulatory failure cannot be deduced from the current lack of new orders for nuclear power plants, but rather from the absence of a rational and expeditious system for licensing and overseeing nuclear operations. The cure for such a regulatory failure is to put the institutions and statutes in order. If utilities, faced with a rational set of regulations and incentives on nuclear power, fail to resume new orders for plants, so be it. The point is that utilities considering nuclear generation now face a tangle of regulations and incentives. Some of them artificially discourage and some artificially encourage a decision to choose nuclear capacity, but all confuse the assessment of true costs and benefits of nuclear capacity.

Nuclear technology is more than "a better way to boil water" for steam electric generation. Since it produces radioactive products, it has the potential to lead to loss of life, health, and environmental quality.

Civilian use of nuclear technology had its origins in the Atomic Energy Act of 1954, which envisioned the Atomic Energy Commission as both promoter and regulator of nuclear applications. The AEC remained as promoter and regulator until the Energy Reorganization Act of 1974 split off the regulatory function to form the NRC. Neither the 1954 act nor the 1974 act contained an explicit consideration of the acceptable degree of risk from nuclear technology; instead, the legislation only suggested that regulators should protect public health and safety. It was as though Congress recognized only the possibilities of having "safe" or "unsafe" nuclear plants, and naturally preferred that they be "safe."

Consideration of the cost of providing nuclear safety would have shown policy makers that it is meaningless to talk about safety except as a matter of degree. There are countless measures that could make a given nuclear installation more or less safe, and research and development can always yield some change in safety technology. But as a plant is made safer and safer, the cost of increasing safety becomes higher. Thus decisions on nuclear safety amount to decisions on how high to climb along the cost curve for safety and how much effort to put into safety research and development—not simple yes-or-no decisions such as "Is the plant safe?" As nuclear safety is pursued further, additional safety improvements become very expensive and the benefits of eliminating the residual risk become small. There would then be an optimal degree of safety somewhat short of zero risk. But how far short of zero risk? On this crucial issue Congress provided not even the most basic qualitative guidance.

Regardless of where the social optimum of nuclear risk is achieved, the amount of risk acceptable to the nuclear firm is likely to be greater than the amount acceptable to society. Indeed, if the nuclear firm had more interest in safety than society has, there would be no rationale for mandatory safety regulation. There are three reasons that nuclear firms, as interested as they are in safety, are still not as motivated to be safe as efficiency requires. (1) As corporations, the firms can lose no more than their net worth, even if society loses much more. (2) Part of the risk is to future generations not adequately represented in current decisions. (3) By statute, the liability of nuclear firms for accident damages is severely limited. Evidence gathered after the Three Mile Island accident, including some not specific to Three Mile Island, indicates that nuclear firms do respond to incentives for safety. This makes the incentives for safety very important in any plan to improve nuclear regulation.

When the NRC took over nuclear regulation in 1975, broad social questions on the acceptability of risk were not any better settled than they had been in previous years. At this point the NRC set out to make nuclear technology safer. Safer than what, or how much safety would be enough, were not questions that were openly addressed. There were, however, pressures on the NRC and its staff not to regulate nuclear power out of business. In view of the risks of the alternatives to nuclear power, it is not clear that regulating nuclear power out of business would lower the aggregate risk to society from energy technologies.

With its complex mission, the NRC would need a clear and coherent internal organization to succeed. It has not had and does not now have such an organization. In many ways, there are two NRCs: the "downtown NRC" in Washington consisting of the commissioners and their staff, and the "suburban NRC" in Maryland consisting of the working program offices that conduct day-to-day regulation. The isolation of the program offices from the commission is at least as great as the physical isolation of the offices implies.

An important part of nuclear regulation is the result of licensing proceedings for specific plants. Licensing proceedings are adjudicatory and adversary in nature, deciding not only technical questions about plant design but also distinctly social questions, such as whether the power to be produced is needed. Opposition to nuclear power often takes the form of intervention into licensing proceedings for specific facilities, so that social questions on the acceptability of risk are decided anew with each case.

After reviewing the mandate and the structure of the NRC, this study evaluated available evidence on the results of nuclear safety regulation. One thing we do know about the state of nuclear safety is that it is uncertain. The worst-case results of a nuclear accident indicate thousands of deaths. But the likelihood of such an accident, or the whole spectrum of lesser accidents, cannot be stated reliably. The AEC-NRC *Reactor Safety Study*, an ambitious technical attempt to estimate the probabilities and consequences of the spectrum of accidents, indicates extremely low probabilities of the most serious accidents. The study has grave difficulties as a basis for policy, however, and following peer review of its technical shortcomings it was withdrawn as an official NRC document.

There is little evidence to confirm the *Reactor Safety Study*'s optimism and several indications that would cast doubt on it. Recent work indicates possible consequences many times higher than the worst consequences reported in the *Reactor Safety Study*. The premiums charged for private nuclear insurance indicate probabilities much higher than *Reactor Safety Study* estimates. And the absence of a catastrophic accident in the limited experience with nuclear power so far can demonstrate a probability of accident no smaller than 1 in 600 per reactor per year. Finally, the Three Mile Island accident is consistent with accident probabilities much higher than those in the *Reactor Safety Study*.

Three questions about nuclear safety have been raised in this study. (1) Is the current level of nuclear safety minimally acceptable? (2) Is the current level of nuclear safety efficient on cost-benefit grounds? (3) Is nuclear safety being achieved cost effectively? The first two questions are subject to great doubt, though the failure of groups investigating the Three Mile Island accident to call for a nuclear moratorium can perhaps be taken as an indication that nuclear safety is at least minimally acceptable. A cost-benefit justification of current or alternative nuclear safety levels is elusive, in view of the unique challenges posed by measurement of the benefit of forestalling a low-probability catastrophe.

A more definitive answer is possible to the third question: the current level of safety is not being achieved cost effectively. Nuclear regulation has at times explicitly violated cost-effectiveness criteria. More often, regulation has been conducted with criteria conflicting with cost effectiveness in mind, such as a need to make regulation visible. Indeed, from the staff emphasis on safety issues to the formalism of licensing, there is a preoccupation with visible regulation at the expense of cost effectiveness.

This study has also tried to discern what accounts for the current state of nuclear regulation. An important element is the inherent difficulty of the regulatory mission. The effects of not knowing the level of safety or the contribution of regulatory measures to safety are pervasive. Every regulatory agency faces uncertainty, but the uncertainties facing the NRC are enormous. So great is the uncertainty that choosing the most conservative parameter from a distribution does not ensure safety, as the distribution of the parameter may itself be uncertain. Nor does compliance with NRC regulations ensure safety, since the NRC has less intimate knowledge of the plant and the operating environment than does the licensee.

Though nuclear regulation will always be difficult because of uncertainties, the following sources of operating difficulties can be changed.

• Failure to make hard decisions on risk. In the absence of its own safety goal or congressional guidance, the NRC has pursued a strategy of undirected risk reduction. Hard decisions on a coherent approach to safety have been delayed.

• Diffuse responsibility for safety. Both within and among the parties involved in the nuclear industry, the responsibility for safety is diffuse. The NRC holds the operating utility responsible for safety, but much in the construction and operation is out of the utility's hands. The nuclear industry suffers from extensive diffusion of the knowledge of how to achieve safety.

• Economic disincentives to cost-effective safety. Previous studies of nuclear power have concluded that the industry has financial incentives not to pursue safety fully and that the industry is willing to take risks beyond what the public would accept. But previous studies have not recognized that disincentives to safety are a direct product of the industry's severely limited liability for accidents under the Price-Anderson Act. Proper liability would make it somebody's business to correct the diffuse responsibility for safety. Even apart from liability, there may be disincentives to safety from other state and federal regulations.

This study identifies three main fronts on which rational regulation should be pursued.

1. The NRC's house must be put in order. The NRC needs a streamlined management structure to replace the dual staffs and multiple layers of management that now exist. So long as the NRC retains its Byzantine structure, managing the agency as chairman will be beyond the ability of any one person. The best way to transform the NRC into an effective agency would be to adopt the recommendation of the President's Commission on the Accident at Three Mile Island and the NRC-commissioned Special Inquiry Group: replace the five-member commission with a single authoritative administrator.

2. Technical issues must be separated from social ones. In current nuclear licensing cases, the set of issues to be decided is too broad and social. Having to decide social issues separately for each facility contributes to the delay and redundancy of licensing. Deciding on the acceptability of risk would best be done by Congress, though it would be almost as good to have Congress give explicit approval or disapproval to the safety goal worked out by the NRC. With a firm mandate for its safety goal, the NRC could devote its resources to determining whether specific proposed licenses or policies were consistent with the safety goal. The risks presented by a given technical system and methods of controlling them are the technical questions that the NRC is uniquely positioned to answer.

These first two recommendations, if implemented, would mean a more coherent NRC focusing on the technical issues raised in licensing and rule-making proceedings. No longer would the NRC be going in ten directions following five separate social issues.

3. Incentives for safety must be corrected. The most obvious way to correct the incentives for nuclear safety is to remove the provisions of the Price-Anderson Act that provide such an extraordinary shield for the industry against damage claims. Coherent arguments for having the Price-Anderson protection were never advanced in the first place. After the fact, a rationale for Price-Anderson could be constructed if it could be shown that insurance markets had failed to provide the correct coverage. In particular, if insurers charged too much for the risk and artificially limited their coverage, then government insurance or a private liability ceiling might be justified. There has been no such demonstration, however, and the experience of Three Mile Island suggests that insurers did not overestimate the potential for damages.

Corrected incentives for safety could move the NRC away from its counterproductive strategy of direct regulation of details of design and operation. With the licensee better motivated to produce safety, NRC regulation could concentrate on the specific areas where licensee self-interest would not be reliable. Even more important over the long term, more normal liability arrangements would effect a high payoff to research and development of safety technology. As it stands, limited liability makes for a low return to such research and development.

All the easy and obvious measures for improving nuclear regulation have been taken. It is time now for legislation to set aright the state of nuclear regulation. The Udall bill of 1980 is a good starting point, in that it has specific measures to separate social from technical decisions and specific measures to begin the job of correcting safety incentives. A streamlining of the NRC under a single administrator and removal of the limit on liability for nuclear accidents would set the stage for more rational regulation.

If the nuclear industry finds that it cannot come up with innovative methods of covering the risk of accidents, and cannot live with the possibility of large damage claims in the event of an accident, then the time for commercial nuclear power has not yet come. The United States should, in this event, accelerate the move toward conservation and alternative energy sources. If, however, the nuclear industry accepts liability for accident damages, and regulation is reformed accordingly, nuclear power can proceed at a pace consistent with meeting the growth of electricity demand.

The worst of all worlds would be to continue with the current state of indecision. Each month that passes narrows the alternatives available, both nuclear and nonnuclear. The nation may end up choosing its energy sources by default.

Rational nuclear regulation is not so ambitious a goal. But in the absence of regulatory reforms and increases in liability, rational nuclear regulation will not be achieved.

SELECTED AEI PUBLICATIONS

Regulation: The AEI Journal on Government and Society, published bimonthly (one year, $18; two years, $34; single copy, $3.50)

Meeting Human Needs: Toward a New Public Philosophy, Jack A. Meyer, ed. (469 pp., paper $13.95, cloth $34.95)

The Regulation of Air Pollutant Emissions from Motor Vehicles, Lawrence J. White (110 pp., paper $4.95, cloth $13.95)

Federal Coal Leasing Policy: Competition in the Energy Industries, Richard S. Gordon (44 pages, $3.25)

Major Regulatory Initiatives during 1980: The Agencies, the Courts, the Congress (74 pp., $3.75)

Health, Safety, and Environmental Regulation: How Effective? John Charles Daly, mod. (26 pp., $3.75)

Occupational Licensure and Regulation, Simon Rottenberg, ed. (354 pp., paper $8.25, cloth $16.25)

The Decline of Service in the Regulated Industries, Andrew S. Carron and Paul W. MacAvoy (73 pp., $4.25)

A Conversation with Alfred E. Kahn (26 pp., $3.25)

Reforming Regulation, Timothy B. Clark, Marvin H. Kosters, and James C. Miller III, eds. (162 pp., paper $6.25, cloth $14.25)

Prices subject to change without notice.